KT-216-266

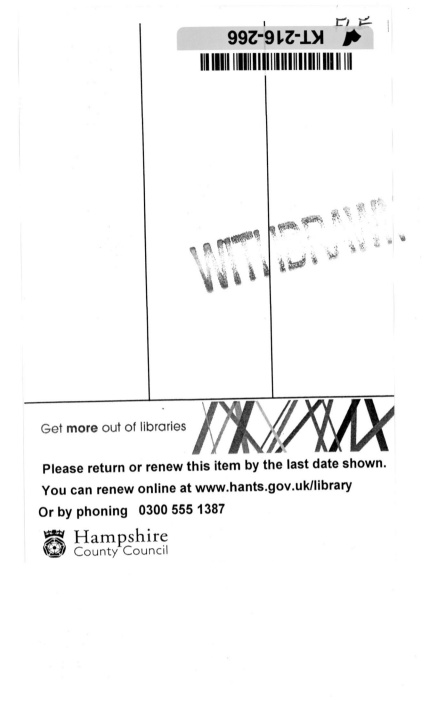

Get **more** out of libraries

Please return or renew this item by the last date shown.

You can renew online at www.hants.gov.uk/library

Or by phoning   0300 555 1387

Hampshire
County Council

C016058555

I wrote this book in the memory of all the mentors, the old guard of naturalists who taught me so much, but more importantly for the new wave, the guardians of the future, to whom we hand over the baton. I include my own little naturalist in this number; Elvie this book is for you.

giving
nature
a home

THE
COMPLETE
NATURALIST

Nick Baker

Foreword by Lee Durrell

B L O O M S B U R Y
LONDON · NEW DELHI · NEW YORK · SYDNEY

giving
nature
a home

The RSPB is the country's largest nature conservation charity, inspiring everyone
to give nature a home so that birds and wildlife can thrive again.

By buying this book you are helping to fund The RSPB's conservation work.

If you would like to know more about the RSPB, visit the website at www.rspb.org.uk,
write to The RSPB, The Lodge, Sandy, Bedfordshire, SG19 2DL, or call 01767 680551.

Bloomsbury Natural History
An imprint of Bloomsbury Publishing Plc

50 Bedford Square
London
WC1B 3DP
UK

1385 Broadway
New York
NY 10018
USA

www.bloomsbury.com

BLOOMSBURY and the Diana logo are trademarks of Bloomsbury Publishing Plc

First published 2015

© Nick Baker, 2015
Photographs © Nick Baker, 2015 except for images credited on pages 341–342
Illustrations © Lizzie Harper, 2015

Nick Baker has asserted his right under the Copyright, Designs and Patents Act, 1988,
to be identified as Author of this work.

All rights reserved. No part of this publication may be reproduced or transmitted in any form or by any
means, electronic or mechanical, including photocopying, recording, or any information storage or
retrieval system, without prior permission in writing from the publishers.

No responsibility for loss caused to any individual or organisation acting on or refraining from action as
a result of the material in this publication can be accepted by Bloomsbury or the author.

A catalogue record for this book is available from the British Library.

Library of Congress Cataloguing-in-Publication data has been applied for.

ISBN: PB: 978-1-4729-1207-7
ePDF: 978-1-4729-2206-9
ePub: 978-1-4729-1208-4

2 4 6 8 10 9 7 5 3 1

Designed in UK by Julie Dando, Fluke Art
Printed in China by C&C Offset Printing Co., Ltd.

MIX
Paper from
responsible sources
FSC® C008047

To find out more about our authors and books visit www.bloomsbury.com. Here you will find extracts,
author interviews, details of forthcoming events and the option to sign up for our newsletters.

Contents

Foreword

Having spent some of the most enjoyable years of my life researching, writing and filming *The Amateur Naturalist* with my late husband, Gerald Durrell, I was delighted to be asked to write this foreword for Nick Baker's *The Complete Naturalist* and eager to see the finished manuscript. How was Nick going to handle the vast subject of natural history, what exciting new ideas and techniques had he come across, would he have had as much fun as Gerry and I did putting the information together? Did he hope, as we hoped, that a generation of readers would take the book to their hearts and learn to cherish nature and become part of the movement to protect it?

My keen anticipation was hugely rewarded, and I was enticed into the natural world all over again by this book. It begins very sensibly with what equipment you need as an amateur naturalist, offering tips as to how to choose and use it, from binoculars and hand lenses to notebooks and clothing. It then leads you logically through the animal and plant kingdoms: mammals, birds, reptiles, amphibians, fish, invertebrates and plants. You get a solid grounding in what to look for and how to observe it, but Nick also grabs your attention with unusual facts – how to tell a right-handed from a left-handed squirrel, for example – punctuated with the occasional hysterically funny personal reminiscence. The activities he suggests are creative and fun, such as recognising bird calls in the dawn chorus, a foolproof method for rearing tadpoles, what you do to preserve a spider's web. The projects are all safe and eco-friendly as well.

Nick has a deft style and quirky sense of humour that brings to life the animals and plants he is writing about. More than anything, he makes you want to spend time outdoors becoming a nature detective. He teaches you how to pick up and interpret signs that reveal an animal's behaviour, and to gather clues that will unravel ecological mysteries.

This is an updated and expanded version of Nick's *The New Amateur Naturalist*, and I am convinced that this book is more important today, ten years later, than ever before. We humans absolutely must understand the natural world. We need to know what its components are and how they come together to make operational ecosystems. We must appreciate how the ecosystems in turn influence each other to make the whole planet tick. Otherwise, the rate at which we change the natural order of things will outpace our ability to correct our environmental mistakes, let along avoid making them in the first place. We can already see our eco-blunders wherever we look – severe floods and droughts and actual and imminent extinctions of animals and plants are examples of the consequences of the human 'footprint' on the planet. But if decisions which impact on the environment are made by people who understand and cherish the natural world, then our tread will be lighter and the planet a more hospitable place for all its inhabitants in the future.

The more youngsters are encouraged to pursue natural history, the more likely it is that they, as the decision-makers of the future, will make the right choices.

Lee Durrell
19 December 2014

Introduction

For as long as I can remember I have been mesmerised by plants and animals, and not just the living, breathing ones. Everything about them, feeding signs and other evidence they leave behind, even their dead bodies can tell us so much about them. But although I have been an amateur naturalist all my life, to this day I continue to learn how and where to look at the living world. That is really what this book is about – using my experiences and the tricks of the trade that I have amassed over the years to gain more insight into the world we live in.

My interest started as soon as I could crawl and pretty soon I was putting my mother through situations no mother can be prepared for: giant silk moths in the wardrobe, tarantulas under the bed and the countless dead animals I would find while out and about and bring home to dismantle at leisure – a form of behaviour my family found particularly disturbing! But to me there was very little difference between wishing to see and understand the internal workings of an animal and my brother pulling a lawn-mower engine to pieces for the same reason.

Through those dark, misunderstood times, a wonderful book called *The Amateur Naturalist* by Gerald and Lee Durrell became my friend, inspiring me to look, investigate and satisfy my natural curiosity. That book was a major influence on my becoming a naturalist, and it was very much the inspiration behind this one. I am deeply grateful to Lee Durrell for providing such a generous foreword to *The Complete Naturalist*.

Things haven't changed much since those early days, despite the fact that I am now a responsible adult with my own house – my home is still stuffed full of natural curios, both living animals and the inevitable collections of debris, skulls, bones and feathers. To me this hands-on approach is totally in keeping with the ethos of this book. You will never really understand something by looking at pictures and writing. Just as you need to stroke a feather to comprehend what an extraordinary combination of form and function it is, you also need to turn a skull over in your hands if you really want to appreciate the beauty of this remarkable collection of bones.

Having said that, you will find a lot more about the living than about the dead in these pages. After all, the living, breathing, breeding natural world is all around us and its influences are felt by all of us, naturalist or not, whether it's the greenfly in your salad, the sticky stuff that gets on your car in the summer or the birdsong that gives a bounce to your step on the way to work or school on a spring morning. Like it or not, we are part of this natural world – it's just that modern life has allowed us to surround ourselves with a cocoon of comfort that isolates us from it.

At its most basic, being an amateur naturalist is simply about enjoying being in touch with our natural surroundings. It's about the joy of observation and discovery, of learning to understand. Some people are put off by the fear that being a naturalist involves learning a lot of science. This is a complete fallacy; to appreciate the miracle of a butterfly emerging from a chrysalis does not require any specialist knowledge. The experience is all that is needed to change the way you look at insects. No PhD required – in fact, a lot of the best and keenest naturalists I know are kids.

On the other hand, you do need an odd collection of personal qualities. You have to be tough enough to be buffeted by the elements and sensitive enough to appreciate the finer points of a Wild Pansy. And even I will admit that it takes a different approach to life to see beauty in a Fox's faeces, not to mention some slick thinking in order to explain yourself to those who catch you looking in the first place!

What I have tried to do in this book is introduce you to the various groups of living things that you are likely to come across in your garden, on country walks and on holiday, and to give you a bit of information on what they are and how they live their lives. I have described a range of useful skills and investigative techniques and explained projects and tricks of the trade that you can try out in the field and when you take your specimens home.

What I really want you to do is get out there, get down on your hands and knees, get your hands dirty, look, learn and enjoy.

What's what and who's who: a quick guide to classification

The way the animal and plant kingdoms are organised may seem a bit baffling at first, with all those long scientific names, but knowing just a bit about it will help you understand which animals and plants are related to each other; that in turn will give you clues to their appearance and behaviour. The kingdoms are the biggest group; the further down the list the more closely related members of a group are, with members of the same species normally the only ones that can mate and produce viable offspring.

Kingdom
Phylum
Class
Order
Family
Genus
Species

'King Philip Came Over For Great Sex' is how the Americans remember the order of classification.

Try not to be daunted by this system. Great naturalists from Linnaeus, Darwin and Wallace until the present day continue to fine-tune, correct and reclassify plants and animals. It is still very much a work in progress as we learn more and more about the natural world and its sweet little mysteries. Keep in mind that this system is intended to make life easier for biologists; just as books are classified in a library, each animal and plant is grouped with other animals and plants with which they have the most similarities.

The scientific names which you will find scattered through this and many other books on the subject are really nothing more than labels. If you always work in the same locality you can get away with using common names, but when you start talking to people in another country, these often fall down and become next to useless. Say something about woodlice to Americans and they will nearly always look baffled, as they know them as sow bugs, but mention their scientific name, and you are on your way to a common understanding.

These names come in two parts: the first is the genus, which may contain similar and closely related species and is always given a capital letter; and the second is the specific name, which is unique to this animal and is written with a small letter.

Codes of conduct

As a naturalist, you have a duty to the natural world around you. I will remind you of these rules again and again in the course of the book, but here is a summary of the most important ones.

- Keep disturbance to a minimum. Never collect more of anything, whether it is a flower or a batch of frog spawn, than you need for your studies, and always release specimens in the same place as you caught them as soon as you have finished looking at them.
- Never handle any living animal unnecessarily. Learn as much as you can from observation alone. If you do have to handle specimens, do it gently and quickly.
- Never make sudden movements. A lot of wild animals are of a nervous disposition and even the smallest of them can scuttle at great speed. Approach them slowly and quietly, from downwind if possible.
- Be extra careful if approaching anything you have reason to believe may be venomous or otherwise harmful to your health. If in doubt, don't do it.
- Do your homework in advance. There are many protected species of plant and animal that you are not allowed to pick or keep without a licence.
- If you are setting traps, bait them with suitable food and water and check them regularly. Many small animals need to eat almost constantly, and you are seriously failing in your obligations if you let them die while they are supposed to be in your care.
- Use a buddy system, particularly if you are going anywhere off the beaten track or in water. Take a reliable friend with you, or at least make sure someone knows where you are and what time you are expecting to be back.
- Take your litter and detritus home or put it in a proper bin. Never, ever, dispose of it at sea.
- While it is always best to observe various species under natural conditions in the wild sometimes it is necessary to keep some animals for a short while in captivity to study. If this is the case it is very important to be aware that some species in some countries are fully protected by law – such as Sand Lizards and Great Crested Newt in the UK – with this in mind where ever you are in the world, it's worth doing you homework and research first, if in doubt contact a local conservation organisation for help and guidance.
- If any animal is taken into captivity – this especially applies to aquatic organisms – it is important that they are kept for the minimal amount of time possible to allow your study and then put back exactly where you found them to minimise the spread of diseases, such as Ranavirus and Chytrid fungus, which are responsible for huge losses in wild amphibian populations. This applies as much to spawn and tadpoles as it does to adult amphibians.
- If you are keeping amphibians in a vivarium, it is essential that both your hands and the vivarium (and vivarium furniture) are thoroughly sterilised and cleaned both before and after the release of your study subjects.

The essential
hardware

Equipment

As a naturalist you will spend a lot of time trying to get close to wildlife while that same wildlife is doing its best to run away from you. Almost every animal you are eager to observe is ready to run at the merest hint of a rustle, cough or hiccup, so it may seem that the odds are stacked against you. But, as you will find throughout this book, there are a few tricks you can employ to redress the balance. The first is to make yourself as invisible as possible when you are close to a subject, and the second is to keep your distance in the first place. From afar you can observe but not interfere, and that's where a pair of binoculars or a telescope come into their own.

Left: A good pair of binoculars is probably one of the most important and useful bits of kit a naturalist could ever own. Invest as much as you can afford, even go secondhand. But having a good pair will pay back dividends.

Right: A bit of a luxury: spotting scopes take you even closer and allow you to really immerse yourself in the details of distant, shy wild lives.

How to hide a human

the art of not being seen

This is a technique that is most useful when you are watching mammals but applies to most other kinds of wildlife too. Remember that when you are out in the field you could bump into a mammal at any moment, and by following these simple rules you can extend the encounter and learn much more.

Blend into the background: wear dark, quiet clothes so that you make as little visual impact and as little noise as possible (see pp.16–18).

Be aware of every noise your body makes: not only obvious things such as footsteps and cracking twigs, but also clothing noise, catching on vegetation and even breathing.

Be conscious of the wind direction: if you suddenly stumble into an exciting situation, it is good to know instinctively where to go. Regularly check even the slightest fairy breath of air movement by dropping a feather, chalk dust or ordinary dust from the ground if the terrain is dry enough. Keep the wind in your face or at least not behind you. For many terrestrial mammals, smell is the most important sense and the one that usually gives you away first. The importance of wind direction cannot be overemphasised; in unfamiliar territory it can literally be a matter of life and death. Sure, it will ruin your day if you come across some deer and the wind direction gives you away, but imagine turning a bush and finding a Black Rhino already mid-charge, because it knew where you were coming from before you even realised it was there! This kind of experience can make for exciting tales, if you survive, but from the point of view of the naturalist who wants to observe without interfering with his subjects, you would have failed. Yes, you would have seen some interesting behaviour, but you would have altered it considerably.

More haste, less speed: move deliberately and expect the unexpected, especially as you move through visual barriers or approach spots where you are likely to find your quarry. Try to think like the animals you are after.

On sighting: move very slowly, using natural cover if possible. Reduce your outline by slowly crouching down to the ground. Keeping low means there is less of your body profile to be seen and you present a non-threatening shape to your quarry.

Never make sudden movements: well, I say this as a general rule, but sometimes I find that, if an animal is distracted for a moment, I can take the opportunity to get into the position I want, quickly. There is a fine line between making yourself comfortable and blowing your chances; only experience will tell you what you can and cannot get away with.

Use your senses: humans are blessed with better noses, ears and eyes than we often give ourselves credit for. Most of us walk about looking but not seeing, hearing but not listening and sniffing but not smelling. Train yourself to use what nature gave you effectively.

Choice of clothing

There is no simple answer to the question of what to wear. It's more a case of what not to wear, and this depends entirely on what you intend to do while wearing it. But there are a number of 'crimes of the cloth' that can be avoided with a little bit of foresight.

Colour really doesn't matter as much as you may think. I have seen people dressed up in military camouflage, looking like extras from an Arnold Schwarzenegger film, with all the latest real tree-print jackets, face nets and gloves, who blow their cover simply by stepping out from behind a bush at the wrong time, moving in the wrong way, breaking cover on the crest of a hill or sneezing. It's really how you move that is critical.

There's no such thing as bad weather, just a bad wardrobe decision! Make sure you're appropriately attired.

But why run the risk of attracting unwanted attention to yourself in the first place? Generally the best colours are dowdy ones that wouldn't look out of place in nature. Personally I like black, as it doesn't look out of place in the street either.

Material: you don't want that rustle in the bushes to be you! Hearing is an important sense for some of the more highly strung animals, acting as an early warning system. So you need to keep noise to a minimum, and that includes your clothes. Waterproof tops or shell layers are the worst culprits. You would be surprised how noisy a pair of nylon trousers or a waterproof jacket can be – they may not be huge on the decibel scale, but they will not only drive you insane as you 'swizz, swizz, swizz' along, but they will also give away your position to any animal in the vicinity. As an alternative, materials such as Ventile, a super-dense cotton weave used to make military immersion suits, may be a bit more expensive but are a worthwhile investment if you are going to spend long hours out of doors. If your budget is small, self-awareness is the key. This isn't a bad discipline for naturalists anyway, so look on it as a kind of training exercise. It is possible to walk quietly in even the noisiest of fabrics, by being aware of what is creating the noise. Usually it is legs rubbing against legs or arms against torso, so try to avoid these movements while you walk.

Zippers or flaps? Zips jingle and rattle with every movement, but can be silenced by sticking down the tag with Blu-tack or duct tape. Space-age technology has given us Velcro, which is great for sealing pockets and zippers, but many a flock of geese or a Pine Marten has been sent flapping or flinging its way into the great yonder by someone diving noisily for a pocketed field guide or Kit Kat. This also brings us to the great Velcro dilemma. Do you tear the surfaces apart quickly and make one short, sharp noise, or slowly, which is a little quieter but with the noise sustained for longer? As always, the situation determines.

Clothing tips

Dress quietly: don't keep loose change in your pockets; the jingle of coins can blow the cover of even the most camouflaged naturalist. Loose Wellington boots are among the worst culprits for making clunking noises. Either get some that fit snugly or wear lace-up boots. Rustly layers are usually waterproof ones. Keep waterproofs tucked away in a bag when you are not using them.

If you are working in the cold, you need good gloves. Mittens are the warmest, although you have to take them off for any tasks requiring dexterity. Some fingerless gloves come with mitten covers and are a great 'best of both worlds' solution.

For flexibility in varying weather conditions, I usually start with a sweat-wicking underlayer (that means that it carries the sweat away from your body and stops you getting uncomfortably hot). Then I build up with multiple lightweight layers and a fleece. A shell layer can be added on top for water- and wind-proofing. When working in the wet, I like waterproof socks. Even if your boots let in water, these socks mean you remain dry right to their tops.

Hats are very useful for colder weather or if you are out at night. Beanies are warm but get wet, brimmed hats are useful for keeping water off your head and away from shoulders but still expose your ears. On hot days a brimmed hat also helps keep the sun off your neck. Baseball caps are useful if you are working in and against the sun. Not only do they keep the sun off your face, but they also act as a visor, meaning you don't have to keep raising your hand to your brow to shield your eyes against the glare.

I like a hat. Beyond the Mick Dundee, Indiana Jones clichés, they serve many purposes beyond the obvious. I've used them for catching snakes and centipedes and transferring frog spawn from a drying out pond.

A versatile piece of lightweight clothing that I find indispensable is the 'buff'. These tubes of material come in a variety of colours and patterns; some are of a thin and stretchy fabric, others are thick and fleecy for winter wear. They can be worn as a kind of draught excluder around your neck, or as a head band, hat, scarf or face mask for keeping out cold air or even breaking up the outline of your face in the field.

Pockets are good, but too many pockets can be bad when it comes to finding what you are looking for. Get into the habit of having special places for certain items, and you will spend less time fumbling.

Binoculars

through those looking glasses

Most of a naturalist's skill lies in observation, and by putting distance between yourself and your subject, you are less likely to influence natural behaviour. Thanks to a Dutch optician who invented the telescope back in the 1600s, we are able to look further afield than we could with the naked eye. With binoculars and telescope, we can effectively draw our subjects closer to us, so that we can see the details without interfering.

As a naturalist you can skimp and bodge and make do with most things, but good binoculars are essential. Fortunately the technology boom has brought the birder's badge of status within the price range of many who would previously have had to choose between buying a car and owning a pair of quality German optics. I'm not an equipment snob, but when it comes to binoculars, accept no compromise. In a nutshell, you get what you pay for, and so always buy the best you can afford. Cheap binoculars rarely deliver. In fact the view through some is so restricted and dull that, despite the magnification, I can honestly say that, if you were to forget the binoculars and use the eyes you were born with, you would see more of the subject! Binoculars should be a pleasure to use. They will become an extension of yourself, and a friend for life, and most importantly, they should be with you and accessible wherever you go.

You mustn't put anything but good glass between you and your subjects. Don't even entertain the plastic cheap thing, you'll be worse off than using your eyes!

So many binoculars: which to choose?

Well, it's horses for courses; first of all, decide what you are going to be using them for and how often. Are they to sit in the glove compartment of your car, or are you going to drag them through the wilds of Outer Mongolia, miles from the nearest lens cloth? Are you going to use them once every leap year, or will they become your life companions, never leaving your bosom? Are you going to hand-hold them for birdwatching or set them up on a tripod to watch crepuscular mammals or scan the ocean for sharks?

Once you have answered these questions, the rest is relatively easy: just keep the following points in mind and remember that you are selecting binoculars for *you*. People are different – what suits one person will not suit another and, because you can be lured into parting with a lot of cash for top models, the wrong choice can be an expensive one.

The price to pay or pay the price: at the risk of preaching, it really is that simple: the more you spend, the better the binoculars, and the better the binoculars, the more you will use them. The better quality the optics, the clearer the image, and because things look great through them, you will use them more often because you want to! They will also be built better, last longer and become friends for life, even heirlooms. Got the message?

What type? There are two main body styles: porroprism, the traditional 'old-school' binocular with an angled body; and roof-prism, the kind with a straight barrel that is fashionable at present. At the cheaper end of the market, porroprism is better as there are fewer reflective surfaces for the light to pass through; start spending more than say £250, and the optically corrected roof prisms rule.

Quality on a budget? Go for a second-hand pair. Good binoculars rarely go wrong, they don't rust and you can tell if they are seriously damaged by simply looking at the lenses and holding them to your eyes – even then a good brand will probably be easily repairable. Once you have worked out exactly what you want, look in the back of local free papers, optical catalogues and birdwatching magazines. This is what I did, and I still own my first pair of Zeiss dialyts.

Magnification: the properties of binoculars are specified by two numbers, such as 8×32, which will be written on them somewhere. The first indicates the magnification and means that the image you see through the binocular will appear that many times closer to you (i.e. eight times closer in this example). Magnification can vary from $4 \times$ to $16 \times$, and the most useful for the naturalist is between $8 \times$ and $10 \times$. For beginners and for those wanting more depth of field, in other words more of the scene in focus, which you would need in dense vegetation, $7 \times$ and $8 \times$ are best; they are also easier to hold steady without technique. For watching raptors and distant birds at sea, $10 \times$ are superb. Anything more than this, and tiny movements from your body, your heartbeat and breathing, combined with environmental factors such as the wind, make the image so shaky that the trade-off is not worth it. Also, the higher the magnification, the duller the image.

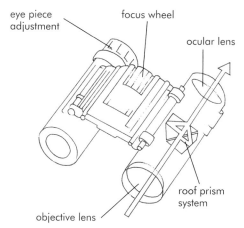

Straight, roof-prism binoculars.

Whalewatching, birdwatching, even trying to find the ice cream van at the end of the beach, binoculars are invaluable when you're out and about. Invest in a good pair.

Generally speaking, the lower the magnification:

- the brighter the image
- the closer the nearest focal point
- the greater the depth of field
- the wider the field of view
- the easier the binoculars are to hold

The higher the magnification:

- the less bright the image
- the narrower the depth of field
- the heavier the binoculars are
- the harder they are to hold still

Stay away from zoom lenses – they are a bit of a gimmick, unless they are built by the higher-end brands, and then they are expensive. Zoom models rarely do what they are supposed to; the quality of the image is inconsistent across the ranges, and so you tend not to use the feature very often; and because the mechanism itself is complicated and fragile, it is more likely to malfunction and need repairing.

Don't get drawn into the brand snobbery that exists; there are many types of binoculars and spotting scopes. Try out as many as you can until you find the right one for you.

Image-stabilising technology, developed by Canon, allows the use of higher magnification in hand-held binoculars without hand shake and is now found in camera lenses, too. Complicated electric-trickery inside the body of the binoculars means steady, high magnifications can be achieved. These binoculars are worth checking out – some people swear by them, though other people complain of a nausea akin to seasickness after extended periods of use.

The letter 'B' after the magnification means that they have push-down or rubber eye caps, so that if you wear glasses, you can use them without reducing your field of view.

Brightness: the second number in the pair gives the diameter of the objective lens. This is the lens through which the light enters at the other end of the binocular from the eyepiece. It may not seem as important as magnification, but it has a huge effect on the quality of the image. The bigger the objective lens, the more light enters the binoculars and the brighter the image. This brightness is important as it determines the detail seen. The size of the binoculars is governed by this second number, not by magnification.

Focus: a maximum of two revolutions of the focus wheel should cover the focus range of the binocular.

Optics: high-density glass (HD) or BAK-4 rather than BK-7 boro-silicate glass may seem an insignificant detail but is the major difference between a dull grey blob and a bird with feathers and identity. It's also the main factor in determining price.

Exit pupil: this is the bright hole you see when you look into the eyepiece from a distance – it represents the light entering the binoculars. The exit pupil is given by dividing the size of the objective lens by the magnification. So for a pair of 8×42 binoculars, the exit pupil is 5.2mm (42/8). Anything above 3.75 mm should cover most naturalists' needs.

Field of view: aim for approximately 120m at 1,000m. The wider the field of view, the easier it is to find your subject. Sometimes the field of view is quoted in degrees, and this refers to the field of view at 1,00 m (about 1,100yd). So if the field of view is quoted as 1°, you will be able to see a range of 17m at a distance of 1,000m.

Glass and prism coatings: go for those that are multi-coated and, in the case of roof prisms, those that have correctional coatings, too.

Quality of build: good-quality binoculars are fairly robust; they may be metal-bodied or even have rubber armour. The initials 'GA' or 'RA' show that there is some kind of armour or protective coating. But build quality can come with a price other than the obvious financial one. The question of weight comes into play, and nobody enjoys hanging a brick around their neck on a piece of string.

Comfort and feel: these are personal things – if the most desirable optics in the world feel wrong in your hand, don't balance well, are a nuisance to use or are to just too heavy, they won't work for you. When buying, try different styles, brands and magnifications until

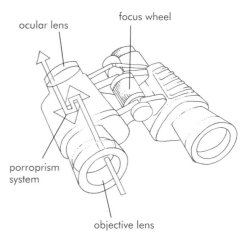

ocular lens   focus wheel

porroprism system

objective lens

Porroprism binoculars.

When it comes to choosing binoculars, the question of weight comes into play, nobody enjoys hanging a brick around their neck on a piece of string.

you find the pair that feels right for you. Choose a weight that will be comfortable hanging around your neck, possibly for hours on end, and a size that will fit your hands but allow your index finger to fall on the focus wheel without stretching.

Indestructibility: another very good reason, in fact the best reason, to splash out. This one word should be used in every binocular catalogue and by every binocular sales rep. A good pair of binoculars is one that you don't have to worry about, that is robust enough to cope with being dragged through bushes and falling off rocks or out of trees. A more expensive pair is also likely to be gas-sealed, which makes it both waterproof and dust-proof. Believe me, this gives you such peace of mind. There is nothing worse than being caught in a downpour and having to worry about your optics getting wet. Mine regularly get a soaking and so far have survived being dropped off a boat into the sea and tangled in the muddy coils of an anaconda.

Tender loving care: another plus for waterproof binoculars is that if they get dusty, sandy or muddy – regular hazards and the kiss of death to the workings and lenses of cheaper designs – you can simply rinse them off under the tap or wash them in mild soapy water and let them dry on the drainer before polishing the lenses with a lens cloth.

Try them out: take your time selecting the binocular for you. Do not allow yourself to be swayed by any sales rep. Try as many pairs as you like. Field centres, observatories and even optical suppliers have open days or will allow you to hold and use their products before you part with your cash. If they don't, go elsewhere – they don't deserve your money.

Protection: once you have selected the exotic optics of your dreams, persuaded the bank to give you that second mortgage and got the pair home, the first thing you should do is get rid of the lens caps. They will be a hassle and a hindrance when you spring for your binoculars in haste to try to identify that bird that's about to dive. In fact, other than for travel, when you need to protect them, your binoculars should never be put away in a case.

Setting up your binoculars

I am frequently horrified when I borrow someone else's binoculars at how badly they are set up or when I mention the adjustment of the diopter and nobody knows what I'm talking about. So here are the two key points to personalising your binoculars and getting the best view on the world.

Get your IPD correct: interpupillary distance is the distance between your eyes. Everyone is different, which is why binoculars come with a hinge. Look at a distant object with the barrels far apart and then move them together until the image is represented by a nice clean circle. Some binoculars have a scale on them for this. If you are in the habit of lending yours, mark your own setting with a pen or a scratch so that you can restore it easily when you get them back.

Adjust your diopter: for most people there is a difference between the focus of the right and left eye. Most binoculars have a function that accommodates this, and getting it right is a beautiful thing, as everything looks much sharper. It feels more relaxing for the eyes, too.

Different designs have different ways of doing this. If your binoculars have a central focusing wheel, shut your right eye and use the wheel to focus the image you see. Then set the right-eye diopter by shutting your left eye and rotating the barrel of the right eyepiece. Make a note or a mark in case you knock it out of alignment. Once your diopter is set, the focus wheel will focus both eyes.

Some brands have a locking diopter ring on the central focusing barrel; it works in much the same way, but is less easily knocked out of position. A few makes have diopter adjustments on both eyepieces – these have to be reset if you change the distance of the object you are viewing.

Other ways to use binoculars

A pair that focus close up can be very handy for watching and identifying larger insects, such as butterflies and dragonflies. Close-focus binoculars allow you to focus on near objects more easily. They can also be used as a magnifying apparatus, enabling you to look at tiny details.

Yet another surprisingly useful quality of a pair of binoculars if you've left your magnifying lens behind.

The field guide
birding with a bible

The two most important bits of equipment a birdwatcher can have are a good pair of binoculars and a good field guide. For me, top American birdwatching advocate Pete Dunne sums up their importance beautifully: 'One confers supernatural intimacy, the other a blueprint to discovery. Together they buy a person passage on a lifelong treasure hunt.'

A field guide has to do exactly what it says on the cover: guide you through the process of distinguishing the species you are looking at from all other possibilities and be practical to use in the field. By this I mean that it has to be easily transportable, preferably small enough to fit in a pocket. For some of the recent works of art that cover countries with a huge diversity of birds, it may be worth considering a pouch or bag designed to fit around your waist by being attached to your belt – a much more comfortable way of carrying a large-format book around than the traditional method of jamming it down the back of your trousers. The latter practice should be restricted to smaller guides and to occasions when you do not have to wear a backpack and when the weather is not so hot that your sweat turns even the more robust publications into papier mâché!

Choosing a field guide is really a matter of personal taste. There is no single one that does everything well – they each have their flaws, biases and layout issues – but you will soon develop your favourites.

To get an idea of some of the best available, check out the *RSPB Handbook of British Birds* or, for North America, *The North American Bird Guide* (2nd edition) by David Sibley. Having said that, I have heard dissatisfied mumblings about both from various bird hides around the world! What it boils down to and the thing to take home is *whatever works for you*. A good book is one that allows *you* to identify the birds that *you* see and, as with all these things, every publisher is always striving to improve, and so even your own favourites may change over time.

A good field guide is organised to enable you to find the bird you want easily – this normally means either in a standard taxonomic order or by visual similarities. To my mind, the most important features are good illustrations that point out noteworthy characteristics and separate a given species from others, especially those species over which confusion often arises in the field. They should also make you aware of any difference between the sexes, or in appearance in flight,

Two birdwatcher's bestsellers, nearly every region of the world now has a fine field guide that you can put in your pocket.

and any other plumage variations such as juvenile, breeding or winter feathers that may throw you off the scent. I avoid photographic guides because I have yet to see one done well. Photographs are restricted in what they can show, and even though a camera doesn't lie, it doesn't always have the flexibility to show all you want to see in the small space allocated to each bird.

I also like a distribution map, ideally on the same page as the identification plates. This is really handy when you are birding in a new and unfamiliar place and have to start from scratch. I will never forget my first trip to Guyana, where I awoke to a dawn chorus unlike any I had ever heard before. As soon as I peered out of the hotel window, it was as if I were five years old again. My birding skills were reduced to excited shouts of, 'Oh look, there's a blue one, and a yellow one and here come five red and green ones!' I had absolutely no idea what I was looking at. Once I had calmed down a bit, my field guide became part of my body, my best friend and an essential organ without which I felt I would perish. Even though there are many pleasures in just looking and watching, the conscious decision to look something up in a book is the beginning of a learning process. To have invested even the tiniest bit of effort in turning a few pages means that the experience somehow sticks better in your head. You become aware of other lookalikes, and you embark on the journey of enlightenment.

One other thing to remember about field guides is not to get too precious about them. If you are scared to use it in case it gets grubby, dogeared or damp, then it isn't working as a field guide. If you are that worried, buy two! Keep one at home all nice and shiny on the bookshelf, and in the other, make notes, draw, tick and add your own observations in the margin. I have seen some birders totally deface their books by tearing out and laminating the colour identification plates to carry around, leaving the text part of the book (the heavy words) behind at home to be checked out later. Others add their own keys and stick on colour references – it really doesn't matter and remember they go out of date.

There are now birding Apps available that can be used on iPhones and Android phones and tablets. They are useful for quick identification, come with bird sounds and some allow easy comparison of similar birds. Apps are portable and often more lightweight than field guides, but you need to buy the host device, which can be expensive. The bird sounds must be used responsibly, see Code of practice p. 57.

Topping it with a telescope

The telescope – the best thing since binoculars! You may think that this is a real luxury in the naturalist's armoury, representing more 'stuff' to cart around, but a good telescope revolutionises the optical and nature experience and, if you have the resources and are serious about your observations, is a very worthwhile investment. If you are worried about size and weight, the new generation of mini spotting scopes is so good that I rarely use my 'Goliath' model any more and opt for the little 'David' version that I can fit into my waist pack.

Buying a telescope requires a certain amount of knowledge, as they come in component form – the body, the eyepiece and the support – and each has some bearing on the other. When choosing a telescope the same overall rules as for binoculars apply, with just a few exceptions:

Scopes come in two configurations – angled (above) and straight (below) – each have their advantages and disadvantages.

Magnification: because you are dealing with much higher magnifications than with binoculars – from 15× to 60× – any instability in your support will be noticeable. In telescopes with a fixed eyepiece, a range of between 15× and 30× is normal. But most now come with interchangeable eyepieces, and so if you have any doubts, you can always buy more than one. For general work, between 20× and 30× is good, but for more distant viewing, you can up the magnification to 40× or even 50× and, though they are not a good idea with binoculars, zoom eyepieces work well with the larger objective lens bodies.

Weight: because of the larger bodies and higher magnification, most telescopes require a larger objective lens. This makes for a heavier device and is the biggest contributor to the telescope being left at home. Smaller, more compact models are a good compromise – try one with a 60mm objective lens and a 20× eyepiece.

Use a rock to stabilise a tripod in the wind.

Colour fringing can be a problem with telescopes. It means that objects appear to be outlined in a coloured halo, especially at higher magnifications. The higher-end models often use extra-low dispersion (ED) glass which eliminates this. Watch out for this disconcerting effect when trying out telescopes.

What type of scope? There are two different body designs to choose from, straight and angled, and both have their uses. The straight body makes it easier to locate your subject – you simply sight it up along the barrel. Some even have a gun-style sight on the outside to help with this. The straight design is also better when you

are sitting down in hides, assuming you can get behind them (some hides now have fixed benches and seats, which can make getting behind your scope difficult) and it tends to be a little cheaper. Angled (with an eyepiece at 45 degrees to the main body) is easier to use if you are tall or if you are viewing birds in trees or in flight, as the angle reduces neck ache. The best feature of this design, though, is that you do not have to have your tripod so high and so it is a more stable set-up. It's also easier if you are sharing your scope with others; people with different heights cope better with an angled scope.

Supports: obviously telescopes can be pricey, especially if you buy more than one eyepiece, but do not skimp when it comes to support. Whether you go for a monopod, tripod or some kind of clamp arrangement, the world's best scope may as well be a cardboard toilet roll tube for all the use it will be with a cheap support.

Photography: if you intend to use your telescope with a camera attachment, it is worth remembering that it was not primarily designed for this function and the quality will be far removed from what you would experience looking through a camera lens of equivalent focal length. Having said that, there is a new generation of digi-scoping technology which utilises the recent advances in digital cameras, with the removal of film and the 'mystery' about what you have managed to take a picture of. This can clarify any doubts about identification and be a handy addition to your field notes, too.

Scopes come in different configurations – this one has a straight eyepiece. Most favour angled nowadays but both have advantages and disadvantages. Have a play.

Using your telescope

Other than knowing which end to look through (it's usually the little end, by the way), the only loosely technical thing is the support the scope sits on. There is an enormous number of different heads and grips, all with different mechanisms. Try out lots of options and choose the one that fits your budget and feels right. With a lower magnification, say 15x–20x, you can use a monopod or lie on your back with your feet in the air and support the scope between your knees – both ways of reducing the amount of weight you cart around with you. But if you can afford it, the perfect combination is a compact scope of the highest quality and a sturdy carbon-fibre tripod – it's what I use all the time.

Carbon fibre tripods are tough but lightweight, excellent for taking out in the field.

Seeing in the dark

the world of image intensification

I remember going Badger-watching on a night as black as pitch, hearing the movements of Badgers all around me and knowing that the moment I turned on my torch I would get a snapshot of Badger life before these highly secretive and jumpy mammals bolted in multiple directions, shattering the moment for all of us. These occasional glimpses were very special in themselves, but the more I did it, the more I would fantasise about how wonderful it would be to have a superhuman ability to see in the dark.

On a moonlit night, when your eyes have become accustomed to the dark, it is possible to see quite well, though details are still a little sketchy. But while I was playing my wishing games the world's military and certain nocturnal hunters were, unbeknown to me, already using the first versions of a technology that could make all my nocturnal dreams come true. Image intensifiers were just around the corner.

The first one I ever looked through was borrowed from my mammology lecturer at university. It was huge, like a bazooka, and seemed just as heavy, like a big, bulky Russian piece of downpipe, with a screw-on screen and a gun grip. Using it or even moving it around late at night made me look like some paramilitary nutcase on nocturnal manoeuvres – something I had a bit of trouble explaining to the police on several occasions!

But whatever the knock-on social and practical difficulties, the moment I turned it on was magic. I may as well have been watching fairies, as I was bathed in the eerie green glow that emanated from the eyepiece and the view of everything in the darkest woodland burned on my retinas. I have hankered after owning my own ever since.

Image intensifier.

Image intensifiers work by gathering ambient light such as moonlight and starlight through the front lens. These packets of light energy, called photons, then enter a photo-cathode that changes light to electrical energy. The energy is amplified by chemical and electric processes and hurled back through a phosphorus screen that turns the electrical signals back into visible – albeit green – light.

Night-vision devices come as first, second, third and fourth generation, a term that refers to the type of light-intensifier tube used. First generation are the most widely available; they also tend to be the cheapest and vary a lot in quality. Some give a reasonable image for the price, but at this level, the technology comes with a whining noise

A volunteer warden uses night vision optics to observe a Manton Bay Osprey pair and their nest during the incubation period at Rutland Water, UK.

and a variable amount of distortion. The quality increases through the generations, with fourth generation being fantastic but well beyond the budget of most naturalists. Night-vision devices of all kinds are rated on two criteria: system light gain – how many times the tube amplifies the available light – and system resolution – how sharp the amplified image appears.

If you decide to buy one of these tubes of magic, many of the rules of choosing binoculars and telescopes apply. But, as the nature of night-vision devices means that it needs to be dark before you can use them, trying them out before you buy can be difficult. But do your best to test some different makes and qualities before you re-mortgage your house in order to afford one! What is most comfortable for you to use is a particularly relevant question, as image intensifiers come in such a variety of sizes and designs, from those that resemble telescopes and binoculars to devices that strap on to your head like a big pair of funky glasses.

One last piece of advice is stay clear of high-street shops selling these products. In my experience they only have them as novelty products and do not know what they are talking about. Always go to a specialist supplier.

I spy with my micro-eye

microscopes

As with binoculars and telescopes, a good microscope is not cheap, but if you are serious about studying the Lilliputian world, you should consider it an investment. It will last for a long time and has very few moving parts to go wrong. So if you have the cash, splash it. Whether you want to examine the internal workings of plants, microscopic animals, the structure of feathers, mammal hairs or even whole insects, microscopes are incredibly useful. I've used one to turn a group of uninterested kids into avid monster spotters! Who would have thought a droplet of greenish pond water with a few dots in it could keep the imagination and sense of discovery going among the PlayStation generation for a couple of hours or more? If you are not convinced, see if you can have a go on one in a laboratory somewhere, and I guarantee you will be converted.

Unless you are a multibillionaire, microscopes come in two forms: binocular and light.

If you plan to work with whole animals such as invertebrates, you want the binocular version. As the name suggests it has two eyepieces, enabling you to view your subject in three dimensions. It is designed to enable you to manipulate your specimen, either by hand or with tools, while focusing on it through the eyepieces – hence its other name of dissecting microscope. Because it is used with relatively large, solid subjects, the magnification is not huge, but most models have either interchangeable eyepieces or lenses of different magnifications mounted on a revolving carousel that give some flexibility, usually between 10× and 60× magnification. Some have platforms with built-in backlights; others have mounted lights that illuminate from above; while with others, you have to provide light from an extra bench-mounted source. If the last is the case, be aware that regular lights are also a source of heat, which your living subjects may not enjoy. The cold light of a fibre-optic lamp is much better, but of course comes at a price. Just keep the health and comfort of your subjects in mind at all times and expose them to bright light for as short a time as possible.

A binocular microscope is perfect for peering into the microcosm of minute invertebrate life.

The light microscope works at much higher magnifications – up to 500×, which is enough to see the internal workings of cells. It is also great for investigating life forms that are normally invisible to the human eye. Stare down a light microscope at a droplet of water from any pond or puddle, and you will be transported into a fantastical world filled with one-eyed aliens, hollow spheres and spaceships.

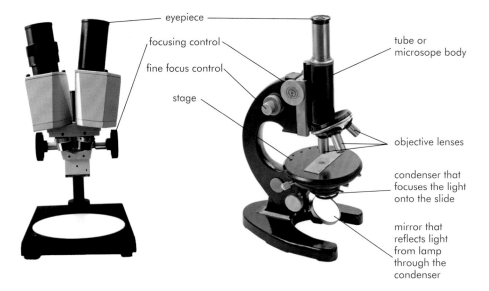

eyepiece

focusing control

fine focus control

stage

tube or
microsope body

objective lenses

condenser that
focuses the light
onto the slide

mirror that
reflects light
from lamp
through the
condenser

binocular microscope   **light microscope**

Because of the nature of these microscopes they require more light
and the subjects have to be semi-transparent to reveal themselves
clearly. So prepare your specimens beforehand: slice plants on a
microtome (the scientist's equivalent of the deli-counter bacon slicer) to
study their structure; squash or restrain other subjects on a microscope
slide; and 'clear' others by treating them in a solution of potassium
hydroxide, which dissolves the soft tissues, allowing the light to shine
through. Staining can be very useful here – it is done with special
biological dyes containing pigments that bond to some compounds in
the specimen but not to others, making certain features stand out.

To keep specimens and preparations, wash them well in water and
then, using a slide with a cavity, position the specimen plus a drop of
gum chloral in the cavity. Seal with a coverslip, wipe away any excess
gum and leave to dry. Larger preserved specimens can be positioned
on a bed of tiny glass beads.

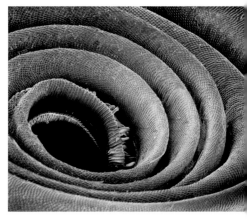

This is what we're talking about.
Although a view such as this, the coiled
proboscis of a Painted Lady Butterfly, is
taken under a very expensive electron
microscope out of the range of a
naturalist's pocket, similarly fascinating
insights can be gained with a light
microscope.

slidebox

forceps

Magnifying lenses, glasses and loupes

Indispensable is the word to describe this small piece of naturalist's kit, whether you are counting the hairs in a cockroach's armpit, scrutinising the meanderings of a red spider mite or counting the stamens in a flower head. The two choices you need to make are what kind and what magnification. And the answers to both these questions very much depend on where and for what you will be using it.

The most basic type of magnifying lens, the type Sherlock Holmes used to use, is not very powerful, but its large field of view makes it handy in the field for observing subjects such as a feeding insect or a nest of ants. Back at base, a similar lens attached to a flexible stem or even a **bench lens** is useful for examining stationary objects, leaving both hands free to manipulate the subject, take notes or draw. Such lenses usually have a magnification of 2× or 3×, not huge but enough to view certain finer details without straining your eyes.

A **watchmaker's lens** is another way of freeing both hands and with a little practice can be gripped in your eye socket for close work – though some people find it takes a bit of getting used to, and your 'eyebrow muscle' may feel tired at the end of an intense session!

By far the most popular and useful hand lens for the naturalist is the **loupe lens**. It is small, folds up into a self-protecting arrangement and is available in a variety of magnifications between 5× and 20×. Anything less than 8× is of little use to the naturalist, while only the most specialist scrutiniser working with the tiniest of details or organisms will ever need more than 15×. If you really cannot decide, you can buy a pocket loupe with multiple lenses. There are many cheap versions out there, but the higher the price the better it will be. Maintenance is no more than a quick rub with a lens cloth or clean tissue from time to time.

It's good to have your lens handy at all times, and so tie a piece of string around it and hang it round your neck. Keep it in its protective case to avoid it being scratched. If you have the choice, buy the kind with an adjustable screw-type pivot, but remember to tighten it regularly – I have had many excellent lenses fall to pieces because I have forgotten to do this.

Hold the loupe in your favoured hand, then rest that hand on the bridge of your nose. Look through it with the opposite eye, focusing by changing the distance between object and loupe – sounds simple, but many struggle to focus by waving the lens around in mid-air.

There are many different types of magnifiers useful to the naturalist, in time you'll find your favourites for the tasks you need to apply them too.

Notebook and notes

I have to be honest with you on this one: I find it hard to write notes when I'm in the field. The moment with a bird or insect is often so fleeting that I get caught up in its magic and forget about jotting down any kind of observation.

Getting into this habit is, however, really, really useful. I cannot emphasise enough how much I have learnt from the few notes I have actually written down. It is very easy to gaze in mindless wonder at the mysterious brown warbler that has just popped out of the bush for the briefest of moments, and not observe anything at all, as your memory will testify when you recall the moment later on.

The sort of awkward questions field guides ask you when you try to identify what you have just seen, like 'What colour were its legs?' 'How long was its supercilium?' and 'Did it have any amount of streaking on its breast?', become so much easier to answer if you get into the discipline of running through the likely points of note while the bird is sitting in front of you, memorising them and jotting them down as soon as you get the chance. It becomes easier with experience, as you get to know the groups of animals you are studying and learn, based on previous difficulties, which points of ID will be most useful to you.

This is something I discovered recently while snorkelling off a coral reef in East Africa. I would see a stunning trigger fish, but on getting out the field guide, I would be confronted with a page containing at least 15 candidates for what I had just observed, all of which could be separated by features much more subtle than those I had noted. Next day, armed with this knowledge, I managed to narrow it down to eight possibles! This continued for several days more until I finally managed to acquire an underwater slate and could take notes on the spot. Then bingo! I nailed it. It took me four days to work out which trigger fish was which, but using a little resourcefulness, I got there in the end.

Probably the single most useful bits of kit are a pocket notebook and a pencil, although modern takes on this such as phone and tablet can do a similar job.

Uses for notebooks

All the great naturalists have made copious notes. It was in the books of such names as Darwin, Bates and Wallace that species were identified and theories on subjects as diverse as speciation, mimicry and evolution came together. Darwin spent most of his life writing up and extrapolating many theories from the notes he made on just a handful of field excursions. Those notebooks still exist and are shedding new light on the biological sciences as we speak. This is one of the greatest uses of notebooks – not only will they be relevant to you as a tool and an exercise to becoming a better and more observant naturalist, but those scruffy scrawlings and scribbles will also provide you with reference and comparisons long after you made them. Just the other day I was referring to a map of a local Badger sett I drew when I was 11 years old. Now, many years later, I can stand in the same spot and see the changes: some holes have long been filled in, trees have fallen down and changed the layout of others, some are brand new and others remain as if no time had passed at all.

Top tips for note-taking

Pencils are better than pens for note-taking: ink can freeze and will run and become illegible if it gets wet on paper. Attaching your pencil to the notebook with a piece of string saves valuable minutes of fumbling in the depths of your pocket or hunting around in the grass for the pencil you 'just put down for a second'.

Buy a reasonably robust notebook, ideally with a waterproof cover and strong binding. For convenience I also use little reporter-type notebooks, the 10 × 6cm (4 × 2½in) sort with a ring binding at the top, and a piece of elastic to hold the pages in place. Having said that, the smaller the notebook the better. The best book for taking notes is the one you have with you in your pocket, not the one with more pages than the *Encyclopaedia Britannica* that you left in the car because you hadn't brought your wheelbarrow with you to transport it!

Keep field notes to a minimum: use your own code and abbreviations to get the information down quickly. Sketches do not need to be something you would want to hang in the Tate Modern, they need only be useful to you. Keeping your lists and notes brief reduces the chances of note-taking becoming an obsessive chore.

Make your notes as soon as you can: the sooner you write it all down, the more you will remember. Regurgitate details such as time, date, weather, wind direction, numbers of flower heads, calls or noises, dimensions, colours, behaviour and anything else you think may be significant. Use drawings, too. They say so much more than words so much more quickly. Birds, winged insects and some mammals simply do not stay still long enough to do a masterpiece, but a few pointers of shape and colour will usually suffice for a positive ID.

You can make notes on phones and take digital photos of almost everything, but nothing hones observation skills and makes details stick in your memory better than actually writing down what you see. Also, ink runs if it gets wet, which is why a pencil can be useful.

Do not worry about scruffiness. The whole point of a field notebook is that it is a tool. If you care about presentation, you can always transfer the information to a master notebook later on. Stick in your original notes if you wish, expand on them, add identifications, etc. You can also embellish your 'master' notebook with specimen items (assuming you are not dealing with a protected species and that collecting is not against the law in any way) such as hairs, leaves, flowers, fruits, seeds and bark rubbings.

Keep your book dry: if you are working in the wet, keep it in a zip-loc bag or even purchase a large bag that you can get your hands into and write in relative shelter. Outdoor, field and forestry stores sell specially designed baffle-type book tents. If this sounds a little tiresome, you can buy waterproof 'write in the rain' books from field and forestry suppliers on the internet. They are more expensive but worth it, as good notes are priceless.

Go digital: if you are technologically minded, you can forsake conventional notes by taking a sound-recording device and a digital camera into the field. You need to be careful not to get these things wet and remember to take spare batteries, memory sticks and sound-recording media with you, but there is definitely room for the techno-naturalist to use new technology to his or her advantage.

A video camera is an expensive piece of recording equipment, but can be incredibly useful for capturing movement, behaviour and sounds.

Sound recording

Another alternative to written notes is the use of a sound-recording device. This can be anything from a cheap dictaphone with a limited amount of recording time to more advanced and flexible media such as mini-discs or mobile phones, which can be saved or the information can be transferred to a computer. Speaking into a microphone is a time-efficient way of recording events and a good option if what you are trying to record is unfolding quickly and in a complex manner or if you don't have a lot of time to write up your notes. Another plus is that you can get a great feel for an environment from the noises it creates; and with certain animal calls, recording them and playing them back can reveal an interesting bit of behaviour.

A mobile phone can be a useful tool in the field, not just for taking photos and notes, but for quickly recording sounds.

Digital cameras

I am a complete convert to digital photography. It is becoming cheaper and cheaper and cameras are getting smaller and smaller, and though for serious photography, only the biggest, most expensive, top-of-the-range stuff is still really an accepted replacement for film, smaller cameras are doing an increasingly good job.

For the naturalist the possibilities of digital photography combined with portable computer technology are limitless. One of the biggest advantages is that, because there is no film, there is less to go wrong mechanically and you do not need to have spare canisters lying around in the bottom of your field bag, getting hot and dusty.

With digital photography comes instant gratification, and because the pictures are free and editable, you can take a picture simply to help you with identification. If you find a mystery butterfly laying eggs on a mystery plant, you can snap the process itself, then the plant and the eggs, and still keep up with the insect, perhaps recording the number of times it stops to lay. The amount of information and the efficiency with which you gather it are greatly increased and you have an accurate and potentially permanent record. Take the pictures home at the end of the day, identify your subject and then decide whether or not you want to keep the information as part of your digital journal.

I have even seen a digital camera used as a magnifying device. I was conducting a field course, part of which involved emptying and recording the species of moth that were being pulled out of a light trap. I was getting in a right fuddle: moths were escaping and I was talking to my students at the same time as I was trying to record the species in my notebook. While I was juggling these tasks, one of my students, a more mature gentleman, was rapidly photographing every moth in sight. I thought no more than that he was taking a very enthusiastic interest in the insects, but later in the evening I found him sitting with his camera, a huge pile of moth books and a notebook. When I asked what he was doing he showed me a comprehensive list of the moths we had seen. He had queries over a few points of identification, and the photographs made it possible to go over this again and even to zoom in close enough to tell the sex of some of the insects. It turned out that this moth enthusiast had bad eyesight and had left his glasses behind, and so rather than slow everyone down in the field, he was snapping away as quickly as he could so that he could identify the insects later and enjoy their details in private.

Small and compact, a pocket digital camera has become a valuable piece of equipment for recording your sightings and findings.

Camera trap

Nets

Most naturalists will find themselves in need of a net at some point – there are some things that you simply can't catch using your hands and arms. Having said that, I believe their use should be kept to a minimum – your eyes are your most important tool, and you should do as much observation as you can without interfering with your subjects. Keep the net for flighty species that tend to disappear before you can say 'Camberwell Beauty', for those that live too high up to be seen properly or for trawling through long grass, ponds or rivers.

Different nets do different jobs – a strong white net with a heavy frame is good for sweeping through vegetation; a lightweight black mesh is better for flying insects; a thick, strong net with big holes for drainage is best for pond work; while putting a jam jar at the bottom of a net will enable you to catch tiny pond animals and plankton. All these are available from specialist shops but are also fun and easy to make yourself.

With butterflies and other flying insects, use the net gently, picking them from vegetation or from behind in flight if possible, and try to avoid hitting your subject with the net rim. Avoid swiping with the net, as this can damage fragile wings. As soon as the insect is in the net, fold it with a quick flick of the wrist, trapping the insect inside. You can now manoeuvre your subject by lifting up the end of the net bag and allowing the insect to fly or crawl up towards the light.

Nets may be great for catching your subject, but they are surprisingly difficult to see details through. For this you need a specialist bit of kit known as a pillbox, which normally comes as a set of varying sizes that sit inside each other like Russian dolls. A pillbox is simply a cardboard pot with a clear bottom, and is fantastically useful for all insects, as they are dark, offer a perch for the insect to grip and, most important of all, being cardboard, breathe, so condensation doesn't build up and the insect doesn't get wet and stick to the sides. If you cannot source purpose-made pillboxes, fashion your own from the cardboard tube that comes in the centre of toilet or kitchen-paper rolls.

To transfer a butterfly or moth from your net, simply cup the pill box over the insect, wait for it to crawl towards the light and slide the lid on. (Another advantage of this design is that it greatly reduces the risk of inadvertently trapping wings and legs, because the insect instinctively heads towards the other end of the box.) Use a small torch to help you pick out tiny details and, as always, as soon as you have finished, release it in the same area that you caught it.

For tips on how to use nets and bits of net etiquette, there's more in the chapter on invertebrates. This is me with my 'old reliable' butterfly net – the frame is at least 39 years old!

Smaller kit

at the sharp end

Tweezers or forceps are useful for manipulating something you don't actually want to touch, like a stinging nettle, or aren't afraid of damaging, like a snail or beetle. Don't use them on small or delicate animals, though – instead go for cocktail sticks or large sewing needles, which give you more control over the pressure you are exerting.

With those frustrating animals that curl up their legs the moment you look at them, try picking them up in a tablespoon or combine this with a small camel-hair paintbrush and use them like a dustpan and brush to dislodge your subjects from a plant or scoop them up off the ground. If you don't want to touch slugs, snails and the like, using two spoons to pick them up gets round the problem.

If you become fascinated by insects and creepy-crawlies generally, you will often find yourself poking and prodding in dark places, and so it can be handy to have something to illuminate the subject. I always carry a pocket torch, and the other device I find useful for bouncing light about is a pocket-sized, travel-type shaving mirror. These are great for looking under ledges and rocks. For nocturnal hunting, use a head torch, keeping both hands free for manipulating any small creatures you may come across.

Pots: you can't have too many

Whatever your passion as a naturalist may be, whether you are collecting seeds, droppings, shells – in fact anything living or dead that needs a bit of protection in your field bag or back at base – you will need pots, and you will soon be recognising the potential of an eclectic range of containers. From ice-cream tubs to the tiny plastic boxes that peppermints come in, you'll find a use for them.

Obviously, if your subject is dead, you can keep it in pretty well anything, but with living creatures, don't forget light and ventilation. Punch a few holes in the lid of a margarine pot, and your spider will be perfectly happy; use an elastic band to hold a piece of netting in place over a take-away container, and caterpillars will crawl around to their heart's content. Use your imagination and think how much your recycling efforts are helping the planet!

Try to collect an assortment of cardboard and plastic pots of different shapes and sizes.

Feathered and flighty

Birds

Birds present the amateur naturalist with some of his or her greatest challenges, for the simple reason that they cannot differentiate between the well-meaning naturalist and the life-threatening predator. So their first instinct is to put as much space as possible between us and them. The same is true of mammals, of course, but birds have the extra disadvantage from the naturalist's point of view of being blessed with wings, making them flighty masters of three dimensions.

Having said that, birds do tend to be visible. Step off a plane into a new and exciting country and the first living things you see when leaving the airport are birds. But catching brief glimpses and getting long, protracted views of them doing what birds do when they are not flying around airports are two totally different ball games! Another important and user-friendly aspect of birds is that many of them vocalise as part of their everyday pattern of behaviour. Generally speaking they are much noisier than mammals, although this behaviour can be seasonal and related to breeding or to environmental conditions.

Right: A murmuration of Starlings is an ornithological spectacle that is hard to beat.

Left: Detail of a Jay's wing feathers.

'Merely glorified reptiles'

That's what the great Victorian biologist Thomas Huxley called birds and, though to the naturalist they present many of the same challenges as do mammals, it may come as a surprise to learn that birds are closely related to reptiles. Some scientists think the two groups have so much in common that they put them in the same class, the Sauropsida.

Of all the terrestrial vertebrates, birds are the most numerous, with something like 9,000 species – there are about twice the numbers of feathered animals as there are furred! In recent years birdwatching has reached the masses and is huge business in Europe and America, with countless societies dedicated to watching and studying them (in the UK, the RSPB has over a million members, while in the US where it is estimated there may be as many as 70 million birdwatchers, the Audubon Society has numerous centres in every state). There are holiday companies that specialise in birdwatching and trade fairs dedicated to all those with an ornithological leaning. The birdwatcher's image has changed, too, and it is a great pleasure now to see people from all walks of life watching birds.

The bits of a bird… useful when describing and trying to identify what you've just seen.

So why are birds so popular?

To claim that they are easy to get to know is inaccurate – they can be some of the most frustrating animals on the planet to study. But as a group they are active and conspicuous, which makes them appealingly accessible. In almost any city in the world you can look out of your window and see a bird. Many are fabulous to look at and have a deep aesthetic appeal; they can also be capable of extraordinary feats of migration and some fascinatingly diverse habits and survival strategies.

Feathers

hundreds of uses for dead keratin

Feathers are the most obvious thing that separates the birds from all other living things. These versatile, disposable, flexible, insulating, lightweight, easily maintained, interlocking, 'scale-like' structures are unique to the birds.

They are also the secret to their success, as bats are the only other masters of true flight, and feathers really do have the edge over the patagium (that flexible thin skin that is the wing of a bat). For a start, if feathers are damaged they can be regrown, which the patagium cannot, and feathers mean that birds don't have to have long, thin, fragile bones, and so their wings are much more robust than a bat's. Feathers make great insulation, which is why birds can maintain their body temperature more efficiently than any mammal – and why we make duvets out of them. A bat's wings, on the other hand, act like giant radiators, which limits where they can live and when they can be active. The fact that a bird's wings (unlike the front limbs of a mammal) are independent of its legs also means that all the other diverse uses to which it puts its feet, such as grasping, preening and manipulating, have been honed to perfection.

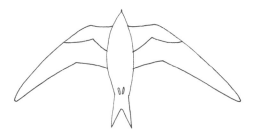

A **swift** has slender, swept-back wings designed for high-speed flight. They are long and narrow to minimise turbulence.

An **albatross** has long, thin wings for great lift and gliding with low effort and maximum efficiency.

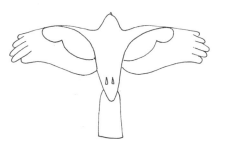

The round wings and long tail of a **hawk** are ideal for manoeuvrability.

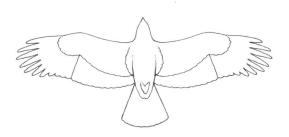

The broad wings and tail of an **eagle**, or **buzzard**, are what is needed for prolonged periods of soaring.

As a boy I was obsessed with feathers. Even today, while I do not go to quite the same lengths to obtain them (I once suspended my unfortunate younger brother by his legs into the Ostrich pen at a local zoo to procure a fine fluffy specimen for me), I still cannot help but bend down and pick them up whenever I see them. Not only are they the most perfectly beautiful example of form and function in the natural world, but many also come in fabulous colours, and for the expert naturalist, every feather tells a tale.

Feathers are made of a flexible protein called keratin, the same stuff as human hair and nails. The reason they are so light for the area they cover is that the shafts are hollow. All the other parts are made up of tiny interlocking strips called barbs, 'zipped' together by little hooks called barbules. If you run your finger backwards along a flight feather, you mess it up, because you have 'unzipped' the barbules. This happens to a lesser extent during everyday wear and tear. If you smooth your finger back up the feather you cause the barbules to 'zip up' again, which is exactly what a bird is doing when it preens itself. Despite their proverbial lightness, feathers make up nearly a quarter of a bird's total weight. They can actually be heavier than its skeleton!

To a naturalist the feather is much more than a bit of windblown fluff; it can tell an awful lot about the previous owner. Different feathers on a bird's body grow from distinct places, or tracts, known as pterylae. When fully developed these feathers take on their own distinctive shape and appearance, and with a little practice, you can identify these.

The apparently perfect flat surface of a feather is made up of a complex series of interlocking 'branches'. The barbs are attached to the central shaft and are locked together by the tiny side branches called barbules.

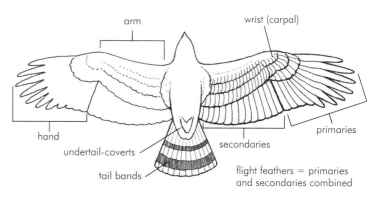

arm

wrist (carpal)

hand

undertail-coverts

tail bands

secondaries

primaries

flight feathers = primaries and secondaries combined

The topography of a typical bird of prey

Super sleuth

The next level of expertise is to be able to identify the difference not only between species but also between sexes, ages and moults. I have a friend who works on prey selection of certain species of raptor such as the Peregrine Falcon (*Falco peregrinus*), and because these birds are messy eaters, if you can retrieve the feathers and meal remains, you can construct a fairly good idea of the bird's menu. Sometimes these dinner remains throw up some real surprises, like new records for species of bird in the area.

Growing feathers

The growth and renewal of feathers, known as a moult, is controlled by hormones and is a fairly energy-expensive operation, requiring a lot of effort and a reduction in efficiency, and so moulting periods are synchronised with other major events in the bird's annual life cycle, such as breeding and migration. An example of this occurs on the predator-free mudflats of the Severn Estuary where in the late summer, after breeding, Shelducks (*Tadorna tadorna*) congregate to moult. This mass moulting is so spectacular that the first time I witnessed it, I thought I was looking at an unseasonable fall of snow – there were literally millions of feathers blowing around in drifts on the muddy shore. Shelducks and other waterfowl become very vulnerable at this stage as their powers of flight are often totally compromised, and so some enter 'eclipse' plumage, becoming very dowdy in appearance to avoid unwanted attention from predatory eyes.

Most birds moult completely two or three times every year for a variety of reasons other than simply to renew the feathers. Some moult into gaudier plumage prior to breeding to display to potential mates. But to continue this beyond the breeding season would only advertise yourself to predators, and so a second moult restores the bird's previous low profile. In other species, feather wear has evolved to occur at a certain pace. The European House Sparrow (*Passer domesticus*), Starling (*Sturnus vulgaris*) and Chaffinch (*Fringilla coelebs*), for instance, all seem to change colour and brightness as winter turns into spring. In reality they do not moult at all – the feather fringes simply wear off and create the illusion of a brighter bird.

Soft downy feathers trap a layer of air and act as insulation. They are used in duvets and down jackets because they are excellent at providing warmth.

Below: This is a primary feather from the right hand wing.

Left: The plumage of the Starling changes colour throughout the year; this is a breeding bird in spring.

Tracking birds

Despite spending a lot of their time above the ground, birds leave a surprising amount of evidence on the ground in the form of nests, feathers and egg fragments. And when they do finally come to earth, they leave tracks. It is a little harder than with mammals to identify individual species by these alone, but once again, combine the physical clues in the footprints and field signs with other information such as the distribution and knowledge of an animal's habitat use, and you can start making good guesses.

To help you narrow down your choices and obtain a positive identification, bird tracks can be divided into three main categories. None of these is linked to phylogenetic relationships; the tracks just coincidentally show a number of shared features.

Ground birds, waterfowl, waders and shorebirds include game birds such as grouse and pheasants, waders such as sandpipers, cranes, gulls and waterfowl such as ducks and geese. All these birds have three toes pointing forwards and one short one pointing backwards. The pattern of the toes tends to be symmetrical. With ducks, geese and gulls, the rearward pointing toe doesn't always register, while with game birds, if it registers, it is as a nail hole. Webbing between the toes of gulls, ducks and geese will show on a soft substrate such as mud and can give you a big clue to the treader's identity, but it doesn't always show up. So always check the shape of the two side toes – if they curve inwards, the track belongs to a web-footed bird.

Perching birds, herons and ibis make up a huge category that includes pretty much every bird capable of gripping a branch – songbirds, birds of prey (excluding owls), pigeons and crows among them. This category encompasses many sizes and lifestyles, and as a result there is large variation in size and a degree of asymmetry between the forward-pointing toes. But the big distinction is that the rear-pointing toe is often nearly as long, if not as long, as the toes pointing forward. Because of their raptorial purpose, a hawk's toes need to be sturdy and well constructed, which is reflected in its footprints.

Woodpeckers, cuckoos, owls, parrots and roadrunners often have two toes pointing backwards or certainly not forwards, giving a rough 'x' or 'k' shape.

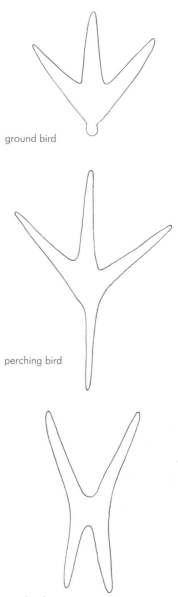

ground bird

perching bird

woodpecker

You can also glean a limited amount of information about the bird's gait from its footprints. Stride length can give you an idea of leg length, and a walking bird will leave an alternating pattern while a hopping bird leaves pairs of tracks next to each other. Shelduck footprints resemble those of any similar-sized waterfowl, but because this bird has a characteristic method of feeding, with a repetitive sideways sweeping of the bill, it leaves a trail of horseshoe-shaped beak tracks as it walks. Dunlin (*Calidris alpina*) work their way across the mud, randomly probing with their bills and leaving a trail of single holes like stitch marks. Snipe (*Gallinago gallinago*), on the other hand, probe with their bills held open, leaving paired 'double stitch marks'.

Who had the eggs for breakfast?

Eggshells often turn up on lawns, or you may stumble upon them out in the field, especially in the spring. The big question is, did it hatch naturally or was it predated? When a chick hatches naturally, it chips away at the shell using its 'egg tooth' and usually makes a neat job near the blunt end. The papery membrane within the egg is preserved and its edge protrudes outside the shell, often curling inward when it dries. The inside of the shell will also be clean. The presence of eggshell doesn't mean there is a nest nearby: once the chicks have hatched the adult often dumps the shells far away so as not to attract predators. A complete egg, particularly if it is pale blue, can be explained by the behaviour of Starlings: one will often nip into another Starling's nest, lay an egg of her own and remove one of the original clutch.

A predated egg, on the other hand, is clearly a predated egg. Whether the egg thief is a bird or a mammal, the job is usually messy; the shell may be split into more than two pieces or simply have a small hole in it; there may also be remains of yolk, egg white or blood (if the embryo was well developed). If the predator was a mammal, the shell may bear witness to this by showing the punctures made by canine teeth. The distance between the marks can also hint at the identity of the predator. The best clue, though, is that the membrane in a predated egg rarely projects beyond the edge of the shell.

The shells of ground-nesting birds such as ducks, gulls and game birds often remain in the nest after hatching. Their remains may be found crushed in the nest or abandoned as the chicks are frog-marched out of the area by the parent birds. A nest like this is worth closer inspection, as it may have been raided (whole clutches can be consumed by predators) and you can use your detecting skills to deduce what has happened.

An egg that has hatched naturally looks cleaner. The membrane only protrudes a little and the broken edge pushes out.

Compared with above, look at the mess made by a Skua feeding on Razorbill and Common Guillemot eggs.

Time to bring up pellets!

We'll talk a lot more about excrement in the Mammals chapter (see p.90), and some of that information applies to birds too. But birds of prey do it differently. They excrete only liquid waste from their vent; their solid matter comes back up the way it came in.

The non-digestible portions of a bird's diet that never make it through the system are regurgitated back up the gullet in the form of a pellet, or cast. Owls are famous pellet-producing birds, but all raptors and members of the heron (Ardeidae), crow (Corvidae) and gull (Laridae) families at some time or another eject these nuggets of indigestible stuff. In the absence of any other evidence, it can be hard to tell which species produced which pellet. But there are a few guidelines that can help you sort this out, starting with the most frequently encountered pellets, those of owls and hawks.

Hawks have much stronger digestive processes than owls and a different way of feeding – they generally tear and pull the flesh from the bone. This means their casts are of a finer texture. All but the biggest bone fragments are dissolved, and in most cases, just the fur and feather fibres make up the bulk.

Goshawk feeding. Its food remains may even include feathers, giving good clues to the identity of its prey.

Finding pellets

The best way to discover these revealing deposits, especially in the case of hawks and owls, is to locate roosts. Owls tend to return to the same places to roost in disused farm buildings, outhouses, tree cavities and even large parkland trees.
They may also make themselves more prominent by being 'whitewashed' by the birds' liquid waste. Hawks also have favourite roost sites, be they fence posts or old trees, and these are your best bet, though the pellets are harder to find as they are produced and 'lost' as the bird goes about its wide-ranging daily business.

With other pellet-producing birds, look around nest sites or under rookeries and heronries. Gull pellets are a common find anywhere on the coast, particularly among tussocky hummocks on cliff tops, but they can turn up pretty much anywhere a gull has stopped. Being scavengers, the contents of gull pellets tend to be very varied and interesting – you can find anything from foil and string to more natural and expected dietary items such as fish bones.

An owl's pellets tend to be much coarser and fibrous as most of their diet consists of smaller prey items that are entirely consumed and sometimes even swallowed whole. Therefore the pellets tend to be more revealing about its diet, and pellet analysis is not only a major part of owl dietary study in its own right, but also a very important way of monitoring the populations of small mammals in an area. These in turn can be sensitive indicators of environmental variations such as land-use change or the arrival of an introduced species.

Pellet analysis

Pull apart an owl pellet, and you will find all manner of bones, not just the big and bulky ones, but also complete skulls of small mammals, small birds, even frogs and lizards. Before embarking on this record the appearance and weight of your pellet.

The pellet of a Barn Owl contains the evidence of feeding from the night before.

Carefully break the pellet into chunks and soak in warm soapy water for a few hours. Tease them out and gently stir or agitate the solution. The heavy stuff such as bones and teeth will sink to the bottom, allowing you to decant and tip away the lighter debris such as fur and feathers. Use a sieve in case you overpour and lose some of the bones, which can be picked up with forceps and returned to the pot. Repeat this process several times until only the bones remain. Transfer them to a shallow dish and sort through them with cocktail sticks or dissecting needles. Use a fine paintbrush to remove any remaining traces of softer tissue.

The parts of the meal that were hard to digest. Bones including skulls, jaws and teeth as well as the fur and feathers of its prey can be found if you poke around with forceps and probes.

Once you have extracted the bones from the pellet, you can, if you like, bleach them (see p.100). Do not leave bones this size in the bleaching solution for long, as they will become even more fragile and the teeth have a tendency to fall out. Display and store your collection of fiddly fragments by gluing them to a piece of black cardboard, and then add the written notes you took before you started.

Birdwatching

Do not mistake simply looking at and identifying birds as birdwatching; for me it means much more – observing their behaviour and learning about their lives and how they interact and work to survive. For example, you might identify eight species among a bunch of waders tottering around on estuary mud, but the true birdwatcher will see how they are spaced out so as not to interfere with each other's easily disturbed prey items; notice that the plovers are involved in a feeding sequence – look, dash, look, peck – while a dunlin randomly stitches its way across the mud. You might even notice three different feeding strategies employed by one species of Oystercatcher!

That is the detail, the fascination of birdwatching. The identity thing, though important, becomes second nature, especially on your home patch. On paper, trying to describe a pigeon and a Peregrine Falcon flying is very much the same, but as soon as you gain experience you will almost instinctively pick up on the GISS of a bird. Though this is often written as 'jizz', it is really an acronym for General Impression Size and Shape and refers to the quick summary, or 'feel', that allows you to identify a species at a glance.

Birding by ear: the art of hearing

Most birds produce a rich range of sounds. Mostly uttered by the bird's syrinx (voice box), these are not just loud, proud territorial claims made mainly by males during the breeding season; the repertoire also includes a complicated array of more subtle avian small talk that is used all year round simply to communicate. There are alarm calls, songs, subsongs, whispered songs, begging calls and contact calls to take into consideration. Some birds also produce sounds other than calls: the 'drumming' display of snipe over a sodden grassland, the irritated 'clack' of an owl's bill or the percussive drumroll of a woodpecker are good examples.

When I started going out on dawn-chorus walks, I was invariably surrounded by people who really knew their stuff. At first I was inspired by the ability of these superior beings to distinguish the subsong of a Blackcap (*Sylvia atricapilla*) from all the other subsongs of a bunch of newly arrived warblers, but soon it became daunting. I retreated into my ignorance and stopped asking questions before slinking off to cry into the pages of my field guide.

It took me years to recover, but here is how I did it. I finally realised that these knowledgeable people were not in possession of a divine

A rich, liquid warbling from the bushes could be any bird to the untrained ear. But combine the sight and sound, in this case, of a Robin, and you will soon be noticing the difference not only between species but between seasons!

Listen for the difference

A noise can be thought of as an audio fingerprint to a species, and it is often easier to make a positive identification of a bird by its call than by its appearance. Take, for instance, the Marsh Tit (*Parus palustris*) and the closely related Willow Tit (*Parus montanus*) – two birds that, even when you have one in each of your hands, still confound many of us! But listen to them call, and you will be in no doubt – if it's sneezy and wheezy, it's a Marsh Tit; if it's more aggressive and scolding, it's a Willow.

Almost impossible to tell apart without a good view, Marsh (above) and Willow Tit are easily identified by their different voices.

gift but had actually learnt their skill over time. And that is the key: *it takes time*. Seeing with your ears is a two-part process. First you have to hear the bird, then you have to make the connection between the noise and the vision. You need to get to the stage where the sound of a song or call instantly takes you back to the moment you first saw it being created. It is a matter of basic association in the same way as a certain piece of music can act as a shortcut to memories of your first kiss or your first pair of binoculars (is he joking, you have to ask yourself?). Collecting these experiences really does boil down to time taken in the field – the more you look, the better you get. Having said that, there are a few 'tricks' that can help you get there faster.

1. Make life easier for yourself: start by working with a few familiar species in your neighbourhood. Get to know their repertoire and move on to other species as you become familiar with those that create the local 'audio wallpaper'. In temperate climes, the best time to do this is at the beginning of the year before all the migrants arrive and confuse things. Teach yourself to listen and hear each voice, then, using all the skills of stealth and patience so key to a naturalist's activities, persist in trying to see and identify the bird making the sound. To home in on different calls and the direction they are coming from, try cupping your hands behind your ears and moving your head about.

2. Sound associations: many bird calls are very distinctive, but to help you remember them, there are various phonetic renderings that can be applied. In Europe the song of the Yellowhammer (*Emberiza citrinella*) can be heard, with a little poetic licence, as 'a little bit of bread and no cheese', the Quail (*Coturnix coturnix*) as 'wet my lips'. With others just the cadence or the sounds can be likened to familiar

Apparently a singing Yellowhammer tells you what it's having for lunch. Its song is said to resemble the phrase 'a little bit of bread and no cheese'.

things: Bobolinks (*Dolichonyx oryzivorus*) have been described as R2D2 shorting out; the common song of the European Chaffinch allegedly has the pace of the footsteps of a fast bowler winding up to bowl; and the Wood Warbler (*Phylloscopus sibilatrix*) sounds like a coin spinning.

3. Expand on the above: mnemonic phrases – those little descriptions you often see in field guides that write the sounds of bird songs and calls as collections of letters – can help once you are at a certain stage of expertise. It is also a neat little trick to draw these sounds! That may seem odd, but when I think of a sound, I often make a mental picture of it. Try drawing things like a wavy line for a rising and falling song, a broken line for short, staccato sounds or an upward curve for a short call that starts low and ends high. This is a technique and a language that will be distinct to you.

4. Keep testing yourself: nobody knows everything; even an experienced birder on his or her own patch will sometimes hear unfamiliar sounds. Never become complacent. Hear a bird, identify it in your head, then go and see if you can catch sight of it and find out if you guessed right. No matter how experienced you are, this is a good way of staying sharp. I often have to relearn the songs of all the migrants that, in the UK, sing only for a month or so every springtime.

5. Use sound references: in the same way as you may use a reference book on your return from the field to help interpret your observation, you can do the same with bird songs and calls. There are a large number of CDs, DVDs, ebooks and Apps that act as audio field guides, helping you to reach positive identifications or remove any niggling doubts about the sounds you have heard.

Male Blackbirds can be very territorially aggressive and will raise a rattling alarm call to chase off other males. Its song is distinctive and tuneful with clear phrases.

In the spring male Willow Warblers often sing from a prominent perch. The song is a series of notes down the scale that can sound quite plaintive.

Mapping territories

The breeding season, particularly the spring for temperate species, is a time for birds to do battle – not with beak and claw but with their voices. Common garden birds are setting up their territories and defending them against their neighbours by shouting all about it.

Now is the perfect time for you to map the territories of the birds in your patch. Simply get up early and look for males singing – they usually choose high perches in the centre of their kingdoms, overlooking as much of their territories as possible. Jot the positions on a map (using different colours for different birds), watch where they fly and try to shadow them, but do not worry if you lose track – you can pick up where you left off the next day. If you see two male birds close to each other, you can assume that each is on the boundary of his territory. Over a few weeks you will start to notice the invisible lines that are the territorial boundaries within which most birds move. Soon you will not only have a unique record of the movements of the birds in your neighbourhood, you will also become a top bird spotter and learn a lot about bird song. If you supply a variety of nesting materials such as bunches of animal hair, wool, string, fluff and even a shallow dish of mud, it won't be long before the locals start visiting; by watching which way they fly off with their beaks full, you can often work out the locations of their nests, too.

Another famous territory exercise is one that should be tried only occasionally. If you repeat it, you will cause the birds a lot of disruption, stress and wasted energy. But it works particularly well with the European Robin and is worth doing once to observe the ferocity of these seemingly harmless little songsters. Just as a male Stickleback responds to the stimulus of a red model fish (see p.156), so the male Robin will attack a model Robin (it doesn't even have to look realistic, just have a red breast) with a fury and boldness that will surprise all who witness it!

Follow a bird gathering material and you'll find the nest – the centre of a bird's world.

Sound recording: catching a song

This technique is not restricted to birds, and much of the following advice can be used to record pretty much any animal vocalisation. As so often, the better the equipment, the more expensive it will be, but to start with cheapish recording gear is fine; you can even make reasonable recordings on your smartphone, especially if you get your hands on a plug-in mic. Essentially you need two bits of kit: a microphone to turn sound into an electrical signal and a recorder to capture and store this information.

Microphones: the biggest obstacle for anyone trying to record wildlife sound (especially bird song) is distance; just halving the distance between you and your subject increases the power of the recording by a factor of four. This problem can be overcome in many cunning ways, such as leaving microphones next to singing posts. But the easiest way around it is a decent microphone. You can think of the microphone as being analogous to the lens of a camera or a pair of binoculars. As with optics it pays for go for quality. To 'catch' a song and minimise background noise, you really need directional microphones (mics designed to record human voices are not as suitable as those used for recording music) or a dish-shaped device called a parabolic reflector, or parabola. Parabolas come in a variety of sizes – the larger the dish, the more the sound is amplified. But it is worth noting that low-frequency sounds such as those uttered by owls and pigeons do not record well if the diameter of the dish is less than the wavelength of the sound – the recording may sound a bit weak and watery as a result. It's a compromise between size and practicality; a dish can be a handful moving through dense undergrowth or in strong winds. Around 50cm (20in) in diameter is pretty good for most purposes.

Recording bird song with a microphone and headphones.

Recorders: the requirements here are simple. Recorders need to be portable – that reads as lightweight – strong, robust and reliable. HiMD Minidisc recorders are good value options, but there is so much to be said for getting the latest solid state recorders – they have less to break. They are, however, quite expensive. The advantage is that they record at what is called a higher bit rate (the best way of understanding this is that it is equivalent to the megapixels of a digital camera) which means more sound information is recorded. The best bet is to start with what you can afford and read up on the various specialised websites before investing in better equipment if the bug takes you.

Evening the odds: attracting birds to you

It's hard being a human when all you want to do is get close to birds and all they want to do is 'vote with their wings' at the slightest perception of a threat. Fortunately there are ways of tricking them into showing their beaks. Become a bird impressionist and you can sometimes 'call them in' so that you can have a closer look.

Interfering with wild birds in this way is controversial and while it was something I would do regularly as a young boy, nowadays the technique is much more widespread, especially with the advent of the smartphone. I'm including this for completeness, but urge any young naturalist to think first and put the welfare of the birds first.

The most universal technique, which works well with lots of small passerine birds throughout the world, is a stylised form of alarm call known as **'pishing'**, producing by blowing air sharply over your tongue while it is squashed between the roof of your mouth and your lower teeth. The noise you are aiming for is a kind of 'pursheeee".

It works best in a sequence of three, repeated at intervals. The birds hear you sounding the alarm and come to see what predator they have missed – behaviour you may have seen when small birds 'mob' magpies, crows, owls and raptors.

When 'pishing' fails, try chipping – a short, sharp, kissing noise between tight lips. This is sometimes enough to make the birds appear, if briefly, out in the open.

When birds are flying overhead and you have no time to reach for your binoculars, try a short 'pew-pew' or a single loud 'pish'; these calls are sometimes enough to cause the flyer to plummet out of the sky to avoid the imaginary predator that you have just warned it of.

'Calling in' by squeaking works equally well for predators such as Foxes, Stoats and Barn Owls.

Make an owl call by blowing between your thumbs into cupped hands.

Put your thumbs together and blow

You may think that, being nocturnal, well-camouflaged, woodland birds, owls would be pretty well impossible to see. But Europe's Tawny Owl is fairly 'birder friendly' as owls go. It is an adaptable bird and, although principally a woodland species, it is tolerant of human disturbance and the only owl to have populations in our cities – in fact, some of my best owling moments have taken place in a car park, watching a Tawny hunt young rats foraging around a litter bin from the vantage point of a No Parking sign.

Calling them in works best in the autumn and winter. For a start the leaves are falling, allowing better views. But also at this time of year there is a lot of shouting in the owl world after dark – the young have been ousted from the territories in which they were raised, and they and other birds are competing for territories that must not only supply food through the coming winter but also have suitable resources such as roosts and nest holes for the late-winter breeding season.

Blowing between your thumbs into an airtight cavity produced by your cupped hands generates a Tawny Owl impression that, though not a patch on the real thing, is often enough to start the blood of a resident male boiling. Do this near where you hear owls calling, and with the use of a torch, you are almost guaranteed to see an owl. At close quarters you will also appreciate that cliché sound of the night, not as the well-known and over-simplified 'hoo-hoo-hooo', but as a throaty affair with wavers and croaks. Play fair, though – once you have set eyes on your bird, stop the impressions and let the owl go back to its business undisturbed.

Probably the most bizarre method is that used to attract the European Nightjar (*Caprimulgus europaeus*) in spring when the birds have just arrived back on their heathland breeding grounds

Don't be deceived by that beautiful face and the duvet-soft feathers of the Barn Owl – this is a serious predator with beak and talons designed to kill.

and are setting up territories. If you approach a site at dusk, just as the birds are getting up for the night, and wave white handkerchiefs in the air whilst clapping your hands together like a morris dancer, you may well lure in a testosterone-laden male bird, also flashing his white wing patches and clapping them together in a bid to challenge you.

This last trick can be much improved (and I have had a nightjar think about perching on my head while I was doing this)

Nightjars generally use their far-carrying frog-like call only in the mating season. Males also clap their wings together as part of the courtship ritual or to warn off rival males. But note that Nightjars are a Schedule One species and as such are protected by the law, so disturbing them at the nest is a serious offence.

by playing a recording of its own call back through speakers; this is known as 'tape luring'. This way you do not even have to attempt to impersonate the bird, and nearly every time I have tried this I have had success. Not just with nightjars, either – many other species are attracted to their own voices.

Code of practice

On the face of it, it may seem fun impersonating or tape-luring a bird, but it's worth considering the fact that you are at best inevitably distracting it from its activities, at worst you are distressing it and other birds in the neighbourhood, and if it is by a nest, the alarmed activity can easily draw the attention of a predator. It is undoubtedly a useful technique among scientists and bird surveyors and while a magical experience can no doubt be generated and a difficult bird might be seen it is worth bearing in mind some codes of conduct. The advent of the same smartphone apps that help us learn bird song has also led to an increased use of tape luring. Normally if bird calls or song are played for a very short time minimal disturbance is caused. But in heavily birded areas such as nature reserves, or with rare and sensitive species, it is generally considered bad practice; also these stressed responses whether territorial or predatory use energy that birds do not always have to spare. While the science is a little confused, common sense should really prevail; if we were talking about individuals occasionally using the technique – then it's not really a problem, but when you've got a large number of birdwatchers visiting a small site or a tiny population and they are all doing the same thing to get a view of the bird, then this is a bit more serious.

I prefer to subscribe to the school of using your field skills to get you a view of the bird behaving naturally. In many countries disturbing birds at the nest is illegal even for common species – in some instances using call playback and tape luring is considered disturbance and to do it you would be breaking the law.

The way to a bird's confidence is through its stomach

When you feed birds in the park or put up a nestbox for them in your back yard, you are providing a necessary resource. This is a good thing and it is even better because it's not a one-way relationship; it makes you feel good for helping out and if you're clever in how and where you position your feeders or nesting boxes you can improve your views and enhance your understanding of their fascinating lives. There are many books and websites dedicated to this vast subject, so once again what I will do here is point you in the right direction and suggest a few of the cheaper and more eco-friendly ways of achieving the desired results.

Feeding first: there is a battery of different feeders available to dispense seeds, nuts, kitchen scraps, suet-based recipes and water to your birds; they vary in design from a piece of board to a top-of-the-range, squirrel-proof affair that comes as part of a huge and burgeoning market in purpose-built, designer bird feeders.

With some inventiveness tins of the kind coffee and baby food come in can be turned into seed feeders by drilling three holes in the sides close to the bottom, then one through the centre of the base. Align this with a hole in a plastic saucer or drip tray, then either screw the whole thing onto a post or suspend it and fill it with seed.

You can make a feeder at home or you can spend as much as you like to attract these animals close up.

Blue Tits are the most common visitors to nut feeders in most parts of Britain.

The Bullfinch feeds on various kinds of seeds. It may nest in trees of larger gardens, and so if you are lucky enough to have a resident pair, it is easy to attract them to a feeder.

~ MAKING A SEED FEEDER ~

You can make your own seed feeder from those horrible plastic milk bottles that seem to have taken over the supermarket shelves recently. I recommend the 1 litre (2 pints) or larger sizes, unless you do not mind refilling them twice a day.

You will need:

- a large plastic bottle
- a knife strong enough to cut through it
- a pen
- twigs or any straight bits of wood, and string or wire to attach them to branches

1 Draw a line on the bottle 1.5cm (½in) from the bottom on the side opposite the handle, parallel to the base, and cut carefully along it. At each end of the incision, cut up about 3cm (1in).

2 Fold the flap you have just created back into the bottle and, hey presto, the bottom of the bottle becomes a feeding tray.

3 Make a couple of perches from the wood and stick them into holes below the feeding tray.

4 Fill from the top with seeds and hang in the garden.

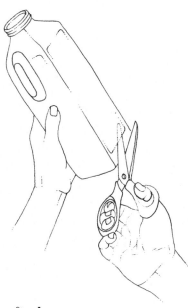

Step **1**

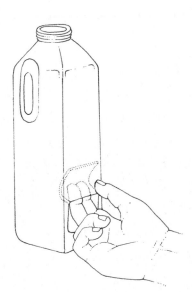

Step **2**

Step **3**

Feathered and flighty 🪶 59

Bird cake or pudding, made from warmed suet mixed with any variety of seeds, fruits, nuts or insects and left to set, can be served up in many ways. Put it in a tin wedged in place between three blocks screwed to your bird table, hang it in pots, cartons or half coconuts, or jam it into large holes drilled in logs – a favourite for woodpeckers.

Mealworms

These are the wriggly larvae of beetles and can be bought in many pet shops, especially those that deal with reptiles and amphibians. There are many mail-order stockists, too, and so a fresh tub will turn up every Wednesday if you so desire. Mealworms can raised in the airing cupboard – keep them in a well-ventilated box with a tight-fitting lid and feed them on oatmeal, bread, biscuits and the like.

A handful of wriggling mealworms may not seem very appetising to you, but Robins find then irresistible. Start by putting a small margarine tub with a few mealworms in it on your lawn to get your local Robin interested. As the days go by, stand out in the garden while it is feeding, inching closer every day. Different Robins have different tolerance levels. So there are no rules about how long getting to within arm's length will take. Just be patient. Keeping low or even lying down may help. Now comes the hard bit – making contact. Offer the mealworm in the same tub, but hold the tub in your outstretched hand. If your Robin is at the right stage of conditioning, it should feed fairly happily. Let him or her settle in to this pattern for a few days, then when he or she seems relaxed, remove the tub and place the worms in your palm. And you should have a hand-tame Robin!

If at any stage you fail, go back a step and keep trying – it is worth it both for the robin, who gets vital food of the right kind, and for you, who will get the rare thrill of contact with a wild bird.

Feeding mealworms to birds can be tricky as they have a tendency to crawl off, something a peanut cannot do. This can be overcome by serving them up in containers with a high enough edge to stop them escaping. Watch for rain, though; drill a drainage hole in the bottom to stop your worms drowning – if they are dead, they stop wriggling and soon go mouldy.

Beware of cheap imitations

When buying mealworms, do not be tempted by cheaper imitations – to the uninitiated, there may not be much difference between maggots and mealworms, but remember that maggots are commercially grown for fishing and may contain chemicals; also they have been brought up on dead-animal material and may contain contaminants that are bad for your birds. Mealworms, being vegetarian, are much more like the caterpillars and grubs that birds feed on in the wild.

Points to remember about feeding

Be patient. If you have just started feeding the birds in your garden, it may take them a while to learn about your service and add you to their daily rounds.

Keep areas under feeders clean. If they are positioned over hard standing, sweep and disinfect it regularly; wash and scrub down your feeding devices as well. Feeding stations, with their high numbers of visitors, are perfect places for disease transfer. Plus the tidier your feeders are, the less likely you are to attract unwanted guests such as rats (though if mammals are your thing, this may be a bonus!).

Use good-quality foods, ideally from a purveyor of seeds and the like aimed at garden birds. Many of the selections you see in the bargain basement have a high concentration of wheat (not a problem if you want to feed pigeons, pheasants and chickens, but not attractive for smaller birds). Peanuts that have been badly stored may contain a fungus called aflatoxin that is lethal to small birds. So buy wisely.

Choose a variety of foods. Not all birds like eating the same things: hummingbirds have specialist requirements and need a sugary liquid; finches go for seeds such as niger and sunflower; tits are crazy about nuts; thrushes like fruit; and woodpeckers love fatty, suet-based stuff. Provide water nearby, so that the birds can drink and wash between meals. This is especially important in the winter if temperatures fall below freezing. Keep the ice away by placing hand warmers or candles below a metal dish of water, regularly pouring in hot water from the kettle, floating a ball in the water or splashing the cash on a heated bird bath (yes, they really do exist).

Put your feeders in an open location but with cover nearby. This means that next door's cat cannot sneak up unawares, but the birds will feel secure knowing there is natural cover into which to dive if a predator such as a hawk shows up. Feed in various locations. Some birds are bold, others shy, so by all means provide food close to the house, but don't forget the nervous ones – put a little at the end of the garden, too. Also bear in mind that different birds feed in different ways; some rarely get onto the bird table but prefer to stay on the ground; others like flat surfaces; some like to hang.

There is a myth that you must stop feeding the birds in the summer/breeding months. This is just not true, as birds are more pressed than ever to meet the demands of their nestlings. What you should do, however, is avoid large food items such as peanuts or bread crusts that can be removed whole, as these, when fed to

Provide a varied menu and you'll get a variety of bird species – Chaffinches (above) are seed eaters and relish a variety of seeds, while Siskins (below) have a finer beak and go crazy for the fiddly little niger seeds. Robins and thrushes have a leaning toward insects – mealworms are a favourite.

nestlings, may cause them to choke and die. In summer keep the food small and soft.

Dependent feeders. Many people worry that the birds in their garden may become so dependent on their feeding that, when they go away, the birds will starve. Well, don't feel you have to cancel your holidays. It is thought that many garden birds treat feeding stations as they do patchy food resources in the wild: as soon as food dries up, they move on to somewhere else in the neighbourhood.

Don't be afraid to experiment. Rather than putting nuts in feeders and crumbs on the lawn, attract a few more finches by collecting teasel seedheads and sprinkling tiny black niger seeds into the natural seed chambers. This makes the seedheads recyclable, whereas in the wild they would be used only once. Try melting fats into some of the seedheads – Blue Tits love it. Alternatively, drill holes in a log, fill them with suet and you have the perfect woodpecker feeder.

Bird cake or pudding: this animal-fat-based food made from warmed suet mixed with any variety of seeds, fruits, nuts or insects and left to set can be served up in many ways. You can provide it in a tin that is simply wedged in place between three blocks screwed to your bird table, it can be hung in pots and cartons as well as half coconuts or jammed into large holes drilled in logs – a favourite for woodpeckers.

A variation on the theme

As well as putting food out for the birds during the winter, how about putting nesting materials out in the spring? When out walking, collect tufts of sheep's wool caught on barbed-wire fences, horses' hair, feathers, dry grass and hay. Add long hairs from the hairbrush and the fluff from the Hoover bag, bundle it up, hang it around the garden or on the bird table and see who comes and takes what for their nests.

Living in a box

nesting sites for birds

The birdbox has been a feature of many gardens for some time, but with the continual squeeze on wild bird nesting places in many parts of the world, due to urban development and drastic changes in land management and farming practice, the back garden is becoming an important sanctuary.

Lack of suitable nest sites holds many bird populations back, and providing them is one way that the amateur naturalist can plough back a little. This is a two-way relationship – you give them nestboxes; in return you get a focus for your studies – and with a bit of techno wizardry, you can elevate the humble tit-box to new heights.

Having birds use a box in your garden allows you to become familiar with intimate goings-on that would usually be hidden in the tangles of the wild. One rather frosty evening, I watched a Wren enter an old nestbox hanging on my garden wall. I was surprised, as up till then no bird had ever condescended to use my home-made box in the breeding season, but I watched with interest, thinking he might be hunting for spiders. Fifteen minutes later, without having noticed him leave, I saw him go in again. After the fifth sighting, I got suspicious and stayed with my eyes glued to the box hole. It soon became apparent that, unless my box had sprung a leak, there were many birds sheltering there; by the time dusk had fallen I had counted 23 Wrens.

Since then I have done a bit of homework and discovered that the record is 61 Wrens in a standard Blue Tit box – the ornithological equivalent of cramming students into a phone box, I guess. The explanation is that the birds are effectively creating one superwren, huddling together and reducing the surface area through which they lose body heat.

Great Tit – a bold bird, likely to be an early coloniser of your nestbox. A nestbox camera (see p.66) gives you a privileged look at your tenants' family life.

Come the beginning of spring, as the birds begin to start calling and warming up for breeding, you will see them beginning to inspect cavities and eventually taking up residence. The real action starts when the eggs inside hatch and the parents' comings and goings increase in response to the demand for food. Just counting these visits and noting the sort of prey they bring in is the key to a much greater understanding of even our most common birds.

Nestboxes are not all box-like – they can be anything from the classic cavity box with a hole to the very latest in woodcrete architecture (a secret combination of wood and concrete that is supposed to have insulating properties and allow the cavity to breathe). They can be a floating raft for Moorhens or a floating beach for terns, a construction the size of a tea chest for owls or a little woven reed or rope ball for Bearded Tits (*Panurus biarmicus*); and these are just those *intended* for nesting. Birds can also be surprisingly good at improvising – every year I hear of a selection of bizarre nest sites that has included overcoat pockets, car exhaust pipes, crash helmets and even a human skull.

The following are a few ideas to be getting on with, but as usual be creative, check out some of the books recommended at the back of this one, and you are sure to come up with a desirable residence for the birds in your garden.

'Woodcrete' is a hard wearing, long-lasting material which has good insulating properties and is ideal for nestboxes.

Sometimes no matter how much effort you go to to provide nesting opportunities, the birds for unknown reasons will not use the boxes. Here a Woodpigeon is using a nest box, but not one put up for it and not in the way intended.

~ MAKING YOUR OWN BIRDBOX ~

This is the most common form, attractive to a whole host of small birds. A hole with a maximum diameter of 32mm (1⅛in) is good for most tits, Tree Sparrows and Nuthatches.

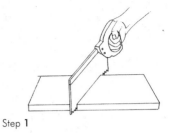

You will need:

- a plank of wood about 15cm (6in) wide, 122.5cm (50in) long and 1.8cm (¾in) deep
- another piece of wood, 40cm (16in) × 10cm (4in), for a batten
- a saw and drill
- nails, screws or glue
- a brass hinge

Step **1**

1 Working on a solid surface such as a DIY workbench, saw a 45cm (18in) length of wood off the plank. Cut this again on the diagonal so that you have two identical pieces, each with one long side measuring 25cm (10in) and the other 20cm (8in). These will make the sides of your box.

2 Drill a hole near the top of the piece that will be the front. (Its size will depend on the size of the bird you want to attract – see p.67.)

Step **2**

3 Cut the remaining length of plank into four pieces, measuring: 25cm (10in) long for the back; 20cm (8in) for the front; 21.5cm (8½in) for the top; and 11cm (4½in) for the floor.

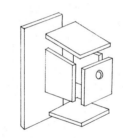

Step **3**

4 Screw, nail or glue the back to the batten, which should stick out a bit at both top and bottom. Then fix the rest of the box together with the longer ends of each side towards the back. Leave the roof till last and fix it on with the hinge. Try to avoid any gaps and you must drill a few holes in the bottom for drainage and ventilation.

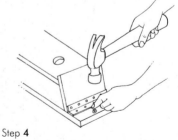

Step **4**

5 Nail through the top and bottom of the batten to fix your box to a tree or wall, high enough to be safe from cats and other predators. See p.67 for advice on positioning.

Step **5**

Specialised birdboxes for fussier birds

If you want to attract Robins, Wagtails and Spotted Flycatchers, make an open-fronted box using the dimensions given above, but with the front piece only about 7.5cm (3in) high. Nail it to a tree or wall in a sheltered spot and partially cover it with climbing plants. Adding a handful of straw or similar material will make it even more attractive. An open-fronted box about twice this size may lure Kestrels, Sparrowhawks or Little Owls.

Treecreepers will visit a box that mimics their preferred natural nesting sites – a crevice in a tree or under a piece of loose bark. Cut two rectangular pieces of wood about 16.5cm (6½in) by 30.5cm (12 in) and two triangular pieces with two sides measuring 16.5cm (6½in) and the other 23.5cm (9¼in). Cut a semi-circular notch about 2.5cm (1in) in diameter out of one long side of each of the rectangles, about 5cm (2in) from the end. Nail the two rectangles together at an angle as shown, with the notches on the outside and near the top, then add the triangular pieces to make a roof and a base. Use mirror plates at the top and bottom to fix it all to a tree, which forms the back of the box, and glue a few strips of bark to the outside so that it blends in with its background.

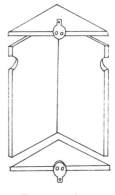

Treecreeper box

If a swift can't find a suitable roof to nest in – quite likely if your area consists of predominantly new houses – it will appreciate a special birdbox. The correct designs for a swift box can be found at www.swift-conservation.org.

Fix the box at least 6m (20ft) off the ground, preferably at roof level, under the eaves, and be patient – it may take the swifts a few years to move in.

Watch the birdie: nestbox cameras

Thanks to technology continuing to make everything smaller and cheaper, it is now possible to buy reasonably priced kits that allow you to view the most secret lives of birds in your nestboxes. A kit consists of a very small infra-red camera with a length of lead that simply plugs into the back of your computer, video or TV. So when there is nothing on the telly except reruns of *Friends*, you can flick over to see what the tits are doing.

The cameras are intended to be fixed into the roof of the nestbox and with some of them you don't even have to do this yourself – the kit comes with its own box. All you have to do is nail it up and plug in, and the birds will do the rest. The lenses are fixed, but because

Location is the single most important factor in the success of your nestbox. Make sure it is high enough off the ground to be safe from predators and facing away from the prevailing wind.

'Location, location, location': positioning the box

You can have the most fabulous, comfortable, centrally heated 'des res' with mealworms on tap, but if it is sited incorrectly, it will house nothing but the beetles you placed in there in the first place. The positioning of your birdboxes is critical; different species have different requirements such as height and relative positioning to other garden features. Robins and Spotted Flycatchers (*Muscicapa striata*) seem

to prefer more open boxes than tits, and sparrows like theirs close to thick bushes; they also choose their nest sites early in the season, and so in the northern hemisphere, put your box up by Christmas if you want to attract sparrows.

The two most important things to take into consideration are shelter from inclement weather, not just rain and wind, but extremes of temperature, too; and safety from predators such as cats. These needs then have to be balanced with your own – how viewable you wish the boxes to be and how easy they are to clean and service (by this I mean an annual scraping out of debris and any necessary external repairs). The time to do any maintenance is during the winter months; this is also the season to reposition the boxes if for some reason you didn't get it right the first time.

the field of view is so wide, the entire contents of the nestbox will be in focus. With the camera high up in the box, the lens remains fairly clean and splatter-free, though when the nestlings start getting cabin fever and realise what their wings are for, the dust does begin to fly. Any muck that gets on the lens can be quickly polished off while mum and dad are away. When your nestlings have moved off, keep watching, because the parents may try for a second family or another pair may move in. If it all goes quiet, you can simply reposition the camera, perhaps in a hedgehog box or on the mammal table.

Once you start 'bugging' your box like this, there is no limit to the possibilities – colour pictures, night vision, microphones, light sensors that allow internal lights to come on and go off. And why limit yourself to the inside of a box? Why not try some of the waterproof units that will allow you CCTV coverage of pretty much every inch of your estate?

With all these hungry mouths to feed, the parent birds, here Great Tits, will welcome any contribution from you – they will eat seeds, suet, nuts and insects.

Hides and blinds

you're outa sight

There is a lot to be said for simply sitting still without moving a muscle, and while this may sound like a great excuse for sloth, there is a point to it. Watching and waiting for something to happen is one of the best things a naturalist can do in the field, and this applies just as much to other wildlife as it does to birds. Obviously you can increase your chances of seeing something interesting by knowing some of the habits of the creature you are trying to watch, but simply doing nothing in the right place at the right time is a technique that has never failed to amaze me. I've had Badgers run into my legs and knock them from under me, I've nearly been stepped on by a Giraffe, I've had a Grizzly Bear eat a fish so close I got an eyeful of milt and I've had a Sparrowhawk pluck a thrush before my very eyes.

The technique is universal, and I remember once having a chat with a naturalist about someone who had named the art of sitting still. The problem is that I cannot remember who it was. Recent conversations seem to suggest a Canadian called Ernest Thompson Seton, though a correspondent of mine who is a member of the Seton Trust has never heard of it but agrees that if it isn't called the Seton technique it should be. So you may well have heard it here first: 'The Seton technique, simply sit and wait!'

Keep your distance and be as quiet as possible.

Sitting and waiting works even better if you cannot be seen by your quarry. There are purpose-built bird hides, or blinds, on many nature reserves now; they can be anything from a draughty shed to a centrally heated, double-glazed monstrosity complete with coffee machine. But the object is the same – they allow you to be human without disturbing the animals you have come to watch. For me the best thing about them is the social scene within: hides are simply the best place to meet other 'birders', from full-blown bird nuts to casual Sunday-afternoon binocular claspers. The hide becomes one big 'superbirder' with as many pairs of eyes as there are people gathered inside, and the more eyes there are looking the less chance a bird has of sneaking past without being seen.

Some hides are the height of luxury – some even have heating and tea making facilities.

Remember hide etiquette, though – keep the noise down, do not forget to share the space and do not, whatever you do, stick your hands out of the slots to point at something. I've seen it happen, and the person who scares off a huge flock of mixed waterfowl by doing this will wish they had wings, too! One other thing – when you leave, shut the viewing 'slots' behind you and don't let the door slam.

At the other end of the hide or blind scale are the ones a versatile naturalist will want to build and position themselves. There were once a couple of nature photographers, the Kearton brothers, who used to get close to animals in many remarkable and bizarre ways such as making model cows and sheep to hide themselves and their cameras in. While there is no limit to how inventive you can be in order to achieve your goal, simpler options are available, and what you opt for depends on your needs.

Probably the most basic form of hide and one you can carry around with you is a bit of military camouflage scrim. Scrim is that netting with various bits and pieces stuck to it, used to break up your outline. Its big advantage is that it is lightweight and you can carry it in your pack. If you stumble across a situation where you need to vanish, simply throw it over yourself or fashion a basic support from the available vegetation and you have a hide. The disadvantages are that it is flimsy, will blow around and doesn't protect you from the elements.

You can achieve similar results with natural materials such as bent branches, string and leafy vegetation or, if you want something more robust and permanent, incorporate waterproof canvas and scrim, add a chair and a flask of hot coffee and you have a set-up that should allow you to outlast even the most patience-testing bird.

The other option is the 'throw money at it' one. There are hides on the market that are practically camouflaged tents and come in a huge range of designs to suit an equally diverse range of budgets.

A hide or blind can be anything from a few branches and fronds to a weatherproof, comfy and collapsible chair hide – you can eve get double ones, so you can take a friend!

The art of not being seen

It's lucky for the naturalist that birds cannot count – even smart birds like Ravens (*Corvus corax*) have trouble with mental arithmetic. If you are planning to spend some time in a hide which has an unprotected approach that means you will be seen by your quarry, rope a friend in to walk you to the door. If you both go in, wait for a minute or two and then your friend goes back; the animals will perceive the threat as having left. You can now relax and get down to watching the animals as they go about their business.

Do your research before building or locating your hide. Get to know the area and how the animals are using it. I once spent a long time creating what I thought was a hide masterpiece, perfectly positioned close to the holt of a Giant

It's all about looking as little like a human as possible.

Otter (*Pteronura brasiliensis*). After waiting for several hours, I became aware of a rumbling noise that appeared to come from below my feet. It wasn't until a strong smell of fish started permeating the hide that I looked out to find that the 'back' door to this otter's holt (actually its front door, as it turned out) opened behind my hide. The rumbling was the otters passing backwards and forwards beneath my feet and the smell of fish came from what they were having for dinner while I was pointing in the other direction!

You don't have to spend money on a hide – making one from materials close to hand is just as effective.

Make a note of the prevailing wind direction of your site and try to pitch your position downwind. This is not as important with birds as it is with mammals, but is still worth taking into account: you may have come for the Oystercatchers, but if an otter was to turn up, you would kick yourself if it sussed you out and ran away.

Prepare yourself for your period of isolation. Take food and a drink of a suitable temperature. Hides can get surprisingly hot or cold, and a drink will help alleviate discomfort. Once inside your hide, do not be lured into a false sense of security. You may be hidden, but any noise you make will pass through the flimsy material of the hide. Avoid rustling clothes, crisp packets and food wrappers. Opt for a quiet sandwich instead. When it's time to go home, always leave as quietly as you arrived.

This simple hide might look like a toilet tent, but it's all that is required to hide the watcher within. The birds get used to objects like this and don't associate the hide with danger.

Birds of a feather

Many bird species find advantage in flocking together for the winter. This whole mass-roosting thing is a bit of an enigma. The first obvious theory is communal thermal regulation – combining lots of little bodies to form one large one with a relatively smaller surface area. This would make sense for a lot of small birds, but if you look at roosts of Pied Wagtails (*Motacilla alba*) and Starlings you will see that the birds are not huddled but very spaced out, and so the theory doesn't seem to work there. Those desperate to conserve warmth, such as Wrens and several members of the tit family, will jam themselves into old nestboxes and crevices, as I described earlier.

So what is going on with the big roosts of birds that aggregate for the night only? An obvious advantage is safety in numbers – lots of birds milling around makes for confusing hunting, and if you watch a flock of Dunlin seethe, shimmer and condense at the attack of a Peregrine, you can understand how this works. This would apply equally to the activities of Sparrowhawks (*Accipiter nisus*) and Tawny Owls, which are regularly seen having a pop at big roosts. Added to the confusion is the fact that, with so many birds to eat, all the predators in a small area are sated quickly, and each individual in a flock stands less chance of being nobbled.

Another theory is that a lot of the bird equivalent of gossiping occurs at a roost, each bird communicating with others if they have split up during the day to feed, as Starlings do. Those returning to the roost with a full crop somehow by appearance or sound pass on the message that they have fed well and that any less successful birds may benefit from joining up with them the next day.

A 'chorus line' performance observed on safari in East Africa – Little Bee-eaters roosting.

Why do Starlings and other birds flock? It's probably as much to do with communicating and temperature regulation as it is about safety in numbers.

Sticks and homes

retro birding

For much of the year woodlands, copses and hedgerows act as living leafy shrouds, obscuring the details of private lives. During the bare winter months, however, things become more transparent. Back lit against a watery grey sky, the trees not only reveal their own profiles but also betray many of the summer's secrets.

Birds' nests are among the things that are revealed. It's taboo to go looking for these in spring and high summer when they are fulfilling their function, but now that they lie vacant – of birds, at least – the amateur naturalist can revel in them.

Dense bundles of twigs and vegetation give away where birds' nests were built, and so if you failed to discover where that Long-tailed Tit (*Aegithalos caudatus*) was heading with all those strands of horsehair and moss back in the summer, now is your chance. If this sounds like a waste of time, bear in mind that some birds return to the same areas year after year or at least select the same habitat or height, and so finding and identifying nests and their positions now can stand you in good stead next year.

At first and to the inexperienced eye, one nest may look very much like the next, just a tangle of sticks, straw and a bit of mud. But with a little practice, a touch of guesswork and the usual healthy helping of dogged persistence, you can soon start linking them to their avian originators.

Distinctive constructions such as the large, loose-domed stick nests of Magpies, the tree-top communities of platforms constructed by Rooks or the solitary efforts of Carrion Crows are relatively easy even from a distance. Slightly more taxing are the similar-sized nests of various garden birds such as Blackbirds (*Turdus merula*) and Song Thrushes (*Turdus philomelos*), but these can be separated on constructional merits. Song Thrushes are unique among British birds in having a hard lining of mud to their nests, while Blackbirds use mud in the construction but actually line the nest with fine grasses. The masterpieces of Wrens look like moss footballs, each with a hole punched in its side, built close to stumps and in dense vegetation.

After some searching you will discover the small cup-shaped nests of finches – often made of finer materials than the thrushes' such as grass, hair, wool and moss. Greenfinches (*Carduelis chloris*) and Chaffinches are less fussy than their relatives about location; their

Song Thrushes make neat, circular nests lined with a smooth coating of rotten wood or dung mixed with saliva.

nests are the ones you are most likely to find in a garden hedge – the Chaffinch's is a rather neater cup-shape that the Greenfinch's. For the other finches, you need to look a little higher, either in the forks of trees close to the trunk or towards the ends of branches.

While nest-watching, look out for stashes of seeds and fruits, as hedgerow nests are often used by squatters such as voles and Wood Mice. And, just as we have a microcosm of life in our own households, so do birds. Take a disused nest home, break it open on a sheet of white paper and watch as pseudoscorpions, spiders and mites come tottering out.

For the observant masterclass of nest-spotters, look for the bored-out nest holes created by woodpeckers. Green (*Picus viridis*) and Great Spotted Woodpeckers (*Dendrocopos major*) have a nest entrance of around 7–8cm (2¾–3in) in diameter, while the Lesser Spotted (*Dendrocopos minor*) has a tiny doorway of about 4 cm (1½in). But there is more to this than the size of the holes. Green Woodpeckers prefer to knock holes in healthy-looking trees with rotten hearts, Great Spotted tend to use trees that are obviously on the way out and Lesser Spotted nest holes are often higher and on the underside of a sloping branch.

It's not just birds' architectural activities that you are likely to come across whilst scanning the lofty levels of a woodland. You may well notice the summer drey of a Grey Squirrel, but the chances are that you wouldn't know that was what it was, as they resemble a hollowed-out crow's nest built high and out on the branches. Larger and much more distinctive is the dense winter drey, also used as a nursery. This is often constructed with leafy twigs, lined with mosses and grass and built close to the trunk where it is less prone to the buffeting of gales.

The Green Woodpecker is not an agile climber; though it nests in trees it is often seen on the ground feeding on ants, its principle food.

Catch the creeper

Another ornithological extra worth looking for requires first identifying a tree. Old parks and churchyards are the best places in the UK to find mature Wellingtonia trees. These unmistakable giant conifers native to California have a soft, deep and fibrous bark. Check this over in daylight, and the chances are you will find small oval depressions in the bark, made more obvious by a trickle of white bird droppings below each one. Return on a cold night, and you will find these plugged with the tiny tawny-streaked bodies of Treecreepers (*Certhia familiaris*). The birds hollow out these customised and insulated snugs, and a single tree can attract birds from all over the neighbourhood seeking sanctuary from the cold; as many as 25 can be seen on one large tree.

Treecreepers have unobtrusive plumage but a distinctive habit of 'creeping' up tree trunks.

The sticky-out bits
watching migrants

From little brown jobs (LBJs) to large white ones, unexpected animals can throw the ornithological world into turmoil at certain seasons. Headlands can be as busy as a bank holiday weekend at Heathrow Airport when the autumn migration is in full swing, what with winter visitors flying in, summer breeders checking out, a few species in transit landing for a refuel and fuselage check and individuals who alight lost, way off their intended course.

Despite the many clues – Swallows gathering on wires, the disappearance of that Spotted Flycatcher that was always in the garden, the emptiness echoing in the shrubs, the lack of calls from Willow Warblers, Common Whitethroats (*Sylvia communis*) and Chiffchaffs (*Phylloscopus collybita*) – the autumn migration is not as obvious as its spring equivalent. In spring the birds are driven by a lustful urgency to set up territories and get a head start on the breeding season; by autumn the pressure is off and the outflux is a gradual one. This apathy is obvious in many streets; while some nests of House Martins (*Delichon urbica*) lie abandoned, their owners already on their way, others still have their entrances stuffed full of the pied 'yippering' heads of the last generation of the year.

Swallows and House Martins belong to the same family and in summer are often seen gathering together on telegraph wires.

Take advantage of vantage points

The knack of beholding the migration spectacle is simply being in the right place at the right time. And the right place is one of the many 'sticky-out' bits: headlands, bills, mulls and peninsulas become the focal points for birds that are passing through, funnelling those moving over land to the shortest over-sea jumps and providing good vantage points from which to spy seabirds passing offshore.

Many birds, such as these White Stork, migrate along well-known flight paths making watching visible migration 'viz migging' possible.

Millions of birds that have visited for the summer breeding season and their offspring are southbound again. Many that have boarded further north, in places such as Iceland, Greenland and Scandinavia, either join us for the winter or use our temperate zones like convenient avian motorway services before continuing south to the Mediterranean and Africa.

If you don't live near moorland and fancy seeing a Ring Ouzel, or want to learn your warblers all in one day, are turned on by rarities such as a Red-breasted Flycatcher (*Ficedula parva*) or simply wish to witness the spectacle of Swallows doing what they are famous for, the time to act is now. Choose your day carefully and according to the weather. For best results, think like a bird that has to conserve its resources; early mornings on days with a inshore wind are best, as any birds wanting to leave land are likely to 'bunch up', waiting for more favourable weather. These conditions also help birds travelling the opposite way and heading for landfall.

Do not be put off by the thought of being surrounded by thousands of unidentifiable LBJs. The beauty of the autumn migration is that, with a bit of luck, you will get plenty of easy 'spots'. A good wind may carry with it the first 'fall' of Redwings (*Turdus iliacus*) and Fieldfares (*T. pilaris*) on the same day that scores of Pied (*Ficedula hypoleuca*) and Spotted Flycatchers, whitethroats, Garden Warblers (*Sylvia borin*), Wheatears (*Oenanthe oenanthe*) and Whinchats (*Saxicola rubetra*) are queuing to check out. In the west it is not unknown for a few treats to turn up – such as Bluethroats (*Luscinia svecica*), Wrynecks (*Jynx torquilla*) or the ungainly bulk of an incoming Woodcock (*Scolopax rusticola*) or two.

Other birds tend to migrate more subtly, small song birds such as this Redstart, make migrations as well, but as loners. Choosing headlands when the weather is favourable, the naturalist can often use this to his or her advantage.

Big game
Mammals

Mammals are the top of the range as far as the amateur naturalist is concerned, simply the best that Mother Nature and millions of years of evolution have come up with. Like us, they are warm-blooded and have highly developed central nervous systems. Many are also blessed with supersenses hundreds if not thousands of times more sensitive than ours – very handy for their everyday survival, but a total pain in the neck for those wishing to study them.

Part of the problem with mammal-watching is that the majority of wild terrestrial mammals are small, highly strung and twitchy, and, like it or not, they see us humans as predators – which is why almost any encounter with one provokes the same reaction: immediate flight. So the naturalist has to find other ways of approaching them and of finding out about their lives.

Below: Badger hair caught in barbed wire.

Left: A male Red Deer with impressive antlers.

Attracting your subjects

It is hardly surprising that so many small mammals are of a nervous disposition. They have the misfortune of being a critical link in a lot of food chains and there is a good chance that pretty much every other animal around that is armed with sharp teeth and claws is going to want a piece of their hide. Many of them have very fast metabolisms, which means they have to spend a lot of their time eating. So if you want to lure them into your garden, you have to provide food and you have to make them feel safe.

The mammal table

The principle of a mammal table is exactly the same as that of a bird table. You put food out and hope to lure some of the shyer furry residents of your backyard within intimate viewing range. It is also perfect for the lazier armchair naturalist who happens to own a set of glass patio doors.

I have seen all sorts of variations on the theme of a mammal table – Foxes, Badgers, deer, opossums, Moose, Pine Martens, Brown Bears and hyenas are all very much up for an easy meal – but here I am going to concentrate on the ubiquitous, but still rather exciting, small mammal fauna.

To set up a successful feeding station, you need to combine the animals' needs with your own wishes. You will want to view with a degree of comfort, and so situate your table near a window. The lower the table, the better, as most small mammals, even those that are good at climbing, will come down to ground level to feed, but not all those at ground level have a head for heights. For the table itself, a piece of smooth wood such as an old tray or a spare plank is ideal.

Providing a certain amount of shelter is important. Covering the platform with chicken wire enables you to see the animals while at the same time dissuading predators such as your local moggy from having a go. Try covered approaches leading from suitable habitats nearby: I have used offcuts from roof gutters and old downpipes with much success. Or you could make the whole table an enclosed box, with one open side facing your glass, for good and secure viewing.

Dawn and dusk are the best times to see mammal activity. These are the times when the day and night shifts change over and most mammals are active, either heading out or heading home. This Wood Mouse is feeding among leaf litter under the cover of darkness.

Mammals such as Wood Mice, voles and rats are the most common visitors to mammal tables – although if you're lucky something larger like a Badger or Fox may put in an appearance.

Somewhere to call home

The small and medium-sized visitors to your table will be more numerous if you encourage them to take up residence near your home. This not only makes it a lot easier for you to monitor their comings and goings, but it is also a way of 'putting a little back', as many natural den sites and habitats in and around human habitation have been lost or damaged by our activities.

This need not involve great cost, heavy construction or sore thumbs. Simply modifying parts of your home or outhouses can make a difference. You could create a bat box or simply leave access points in your own eaves and allow the local bats to occupy existing cavities; the amount of effort you put into this is entirely up to you.

Just as with birds, many mammals appreciate a little help on the housing market; this is a bat box.

Remember also that these animals are very sensitive to noise and disturbance, and so keep this down and put red filters over your torch or any others lights (most nocturnal animals are less sensitive to light in the red end of the spectrum). Bait the table with a variety of food such as grain, cheese, cat food, mealworms and fruit.

Tips for the mammal-watcher

Bearing in mind the natural nervousness of most small mammals (and quite a lot of large ones, come to that) when faced with a member of the human race, there are a few commonsense rules you should bear in mind when attempting to watch them.

There is an inverse ratio between the number and length of sightings made and the number of people in your party. The fewer observers, the more you will observe.

Keep the wind in your face and be aware of natural cover, whether it is vegetation or a bend in the track. Position yourself in such a way as to minimise the likelihood of a surprise encounter. Always expect an animal to appear, act accordingly and be rewarded.

If hides are available, make use of them, remembering to consider the needs of other mammal-watchers too. Alternatively, build your own (see p.70).

If spotted, try to look as unhuman as you can. Get down low and move slowly and inconsistently: take a few steps, wait, take a single step, wait, then continue. You are less likely to be thought of as a potential predator if you don't behave like one.

Your best chance of seeing shy nocturnal mammals such as Pine Marten is during the twilight of dusk and dawn.

Reading the signs
the art of seeing

Encouraging wildlife into your garden is only the first step towards learning more about it. The keen naturalist is also going to want to get out there and study the signs the animals leave as they get on with their lives.

Why track animals? Funnily enough, the answer isn't necessarily to find the animal that made the tracks. Learning how to track is like learning to read all over again. And just as with reading, the aim is not to get to the end of the story as fast as you can but to 'experience' the words.

Tracking an animal is not just a way of getting close to its physical presence but a way of finding out what makes it do what it does. When you begin to understand an animal's ways, you are better able to predict where it is going to be, which leads to a higher chance of actually seeing it, because by using your knowledge you can position yourself accordingly. Any fool can stop and look, but not everyone can stop and look in the right place.

So first you have to learn to recognise 'field signs'. Look close enough and you will find every square metre of a habitat has something to tell you about the animals that live in it or move through it. Some of the signs are very subtle and tell you things in an easily missed 'whisper' – a single hair on a fence post, say. Others are more obvious: uprooted trees with branches twisted off scream 'Elephants were here!'

Interpreting field signs is vital when working with timid and secretive animals, as it is often the only way we can piece together their private lives without trapping them. I know somebody who did a doctorate on the behaviour of Otters in an English river and during the three years of the study never set eyes on the elusive beasts themselves.

There is so much you can tell from a track: not just what made it, but when, what they were doing, how they were moving and sometimes even individual numbers.

So the nature detective spends a lot of time looking at the ground. It never fails to amaze me to see early in the morning the trails left by nocturnal animals. Most are made by domestic animals, but look closely and you will soon start noticing other trails. A lot depends on the substrate where you are looking: to find good tracks you need the soft stuff. Mud and sand have the advantage of being available in most habitats in most seasons.

Let's start with a couple of definitions. It may seem a bit boring, but if you are working with others or trying to interpret someone else's notes, it makes sense to ensure that you are all talking about the same thing.

An experienced naturalist can look at a trail in snow and recognise the 'maker' simply by the pattern and rhythm of the trail.

So, a **track** is specifically the impression made by the feet (or any part of the body involved in locomotion). A **trail** is the overall impression made by an animal's passing by. It can indicate size, gait and the speed at which the animal was travelling. The term trail also refers to marks made by other parts of the animal's body, such as tail drag, and its physical effects on the surrounding environment, such as stripped bark or trampled vegetation.

This footprint of a European Lynx has been photographed with a birch leaf to give a sense of scale.

Tracks

The track, or footprint, of an animal is a good starting point for practising your detecting skills. Roughly speaking tracks can be separated into four categories:

- **birds**;
- **hoofed mammals** such as deer, sheep and horses;
- **animals with 'pads and toes'**, which can be anything from a bear to a mouse;
- **everything else,** from reptiles to insects – after all, even the lowliest worm can leave a track if the conditions are right.

So, for the purposes of tracking, mammals fall into two main categories: those whose limbs terminate in hooves, or cleaves, and those that have paws, with pads and toes.

When mammals evolved 200 million years ago, they started off with a basic, archetypal foot plan that can be likened to the human hand, with five digits. The digits are conventionally numbered one to five, with number one, the inner toe, always being the shortest, analogous to the human thumb. This is relevant to the tracking naturalist because the world is still walked upon by animals with primitive foot layouts – flat-footed or **plantigrade** creatures include insectivores such as hedgehogs and shrews; bears; mustelids (the group to which weasels, badgers, skunks and otters belong); and primates.

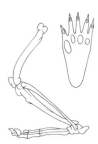

Skeletal leg and foot of a squirrel and its footprint: a typical plantigrade print.

This may sound like an unnecessary amount of detail, but it is just the sort of thing you need to know if you are to interpret field signs to the full. If you find a track showing five toes, you can start making guesses about what sort of animal made it, and if you can identify the inner toe, you can tell from its position whether the track is that of a right or a left foot.

Now the problem with being flat-footed is that you have a lot of resistance – every time the foot goes down it acts as a brake, wasting valuable energy. So mammals that needed to travel fast had to adapt. Over time, the weight of the body moved forward, the legs became longer and mammals started to run on their toes. One of the accompanying changes was that the number of toes was reduced, and the toe to go was our friend digit number one. Look at the footprint of a dog or cat, and you will see it registers only four toes on each foot. In such **digitigrade** animals, the remaining four toes became larger and if there is a short toe it is now the outer one, digit five.

Skeletal leg and foot of a dog and its footprint: a typical digitigrade print.

This process continued with the further reduction of toes in **unguligrade** animals. Deer and sheep produce footprints that register two toes per foot; these cleaves, or slots, correspond to two hooves representing digits three and four. The most extreme toe

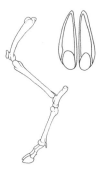

Skeletal leg and foot of a deer and its footprint: a typical unguligrade print.

Padded feet

The pads of a mammal's feet can also help with identification. Next time you are curled up on the sofa with the household cat or dog, check out its feet. The undersides have tough leathery pads that act as springy cushions and are reinforced with protective ligaments. Below the tip of each toe is a corresponding toe pad, then behind these there can be various arrangements of other pads. In some species these blend together to form the central or hind pad (also called the palm or heel pad). The footprint of a human or a Badger shows this well.

Badger paw.

reduction is found in the equines, or horses. A horse stands on the tip of digit three; all the other toes have disappeared.

In some animals, the pads on the feet provide individual signposts. If I'm going somewhere where I'm likely to pick up track, I usually take enough equipment with me to make casts of four or five dog-sized prints – this is a reasonable amount to carry on a day trip. But if there is a chance you will find *big* prints, you need *big* quantities as well as *big* vessels to mix or melt in. The example that springs to mind is my quest for Polar Bear prints in the Canadian tundra. These huge predators leave deep, dinner platter-sized pugmarks, and I thought a print of one of them would be the ultimate addition to the collection.

It was this vision that drove me to the madness and inconvenience of lugging a 15kg (33lb) sack of plaster and bucket with me wherever I went. We were travelling everywhere by helicopter and the sack kept springing leaks and puffing forth the fine white powder, enough to test the patience of any pilot, especially one who clearly took a lot of pride in the interior of his helicopter. On top of this we were working as part of a film crew, and my bulky bag was simply getting in the way. When I was carrying it in my backpack it nearly killed me. It all seemed as if it was going to be worth it when, much sweating, puffing and panting later, I found my dream footprint beautifully registered in some frozen mud. But I was in the sub-arctic in early winter, and I had forgotten one rather obvious fact – that all the available water was frozen solid and I had nothing to mix my plaster with.

As so often, a bit of planning and research would have gone a long way to avoid much discomfort and suffering.

Polar Bear tracks in the snow, unmistakable in size, shape and gait – a textbook plantigrade walker.

Trail patterns

Now let's look beyond individual tracks and consider the overall impression. Even if you are having trouble reading the tracks, it doesn't mean you can't identify the animal.

The footprints of many wild canines are very similar to those of domestic dogs; but look up from the tracks, look at the way the animal has moved and you can learn to read its trail. This is the masterclass of tracking, the difference between being an Inspector Clouseau and a Sherlock Holmes.

Mammals' gaits are all variations on four themes: walking, trotting, running and bounding or hopping. Not every mammal uses all four – rabbits and squirrels, for example, never walk; even when moving slowly they hop. In the field, look out for any distinctions between fore- and hindfeet; many animals have them.

This is a full set of squirrel tracks; you can see quite clearly the different size and shape of the forefeet and the back feet from this perfect impression in the snow.

Walking shows as a pattern of alternating footfalls. The left hindfoot moves first, then the left forefoot, then the right hind, then the right fore, before the process repeats. The spacing between the footfalls depends on the speed the animal is moving and also the type of animal. At one extreme the front and rear tracks superimpose totally, known as perfect registration (seen in some deer); sometimes the registration is imperfect and the hindfoot and forefoot overlap partially, as with a walking Badger. So the trail pattern appears as two parallel lines of tracks.

With **running** there are no alternating movements. When an animal takes off with a burst of speed, both front legs leave the ground, then both hind legs, and for a moment in the stride all feet will be off the ground – the animal is literally flying along. The feet land not simultaneously but offset from each other. The rear legs are used for push off and their impressions are left well in advance of those of the forefeet. When an animal runs, the front and rear track never register and often there is fine evidence of the toes splaying.

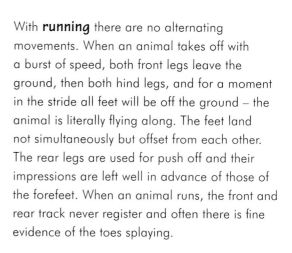

Trotting is a very energy-efficient way for an animal to move. Alternate feet leave the ground at the same time, e.g. right forefoot and left hindfoot. As the speed increases, so does the length of the stride, while the width of the trail decreases.

A **bounding** animal is also airborne at some point in its stride. It pushes off with its hind limbs and lands on its fore, then takes off again on its front feet before the rear legs have had a chance to land. The tracks show the hindfeet in front of the fore.

THE PERFECT PRINT
~ making a good impression ~

If you want a permanent record of an animal's footprint, making a plaster cast is fun. It's also easier and more accurate than drawing and is the best way to compare and contrast different tracks. On top of that, at the end of the day you have a trophy, which counts for a lot when working with mammals – usually all you have is the experience and a story to tell.

Do not necessarily restrict yourself to individual or perfect footprints; different gaits and substrates leave different degrees of registration. The track left by a running animal is very different from that made by the same animal standing still; the same applies to hard or soft ground. Take a variety of casts to illustrate these differences. With small creatures, try casting a number of tracks together, to record the gait.

You will need:

- plaster of Paris/wax beads or chips
- a bottle of water (if using plaster)
- a piece of card
- a paperclip
- a container for mixing the plaster (I usually use a couple of plastic bottles with the tops cut off) or a tin and a small stove for melting the wax
- a soft brush and a mixing stick
- an old knife that you don't mind digging with
- newspaper or other packing material to protect the cast when complete

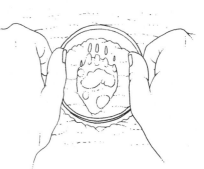

Step 1

1 Once you have found a clear footprint that you wish to keep or identify back at base, make a mould around it. If the print is the right size, you can trim a ring from a plastic bottle; otherwise fold a strip of card round the print and hold it in place with a paperclip. Or, if the substrate is something suitable such as sand or mud, heap this up around the print. Whatever method you use, make the mould deep enough to contain at least 2.5cm (1in) of plaster – any less and the resulting cast may be brittle.

2 Mix the plaster of Paris quickly with enough water to make a treacly consistency, stir to remove any lumps and pour it evenly into the mould. Tap the mould gently a few times to release any air bubbles and wait for 10–15 minutes

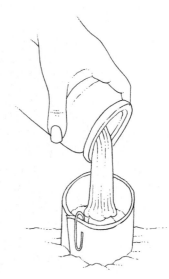

Step 2

for the plaster to harden. The time this takes varies according to the consistency of the mix, the depth of the mould and the temperature. If you are using wax, the same principles apply, but you need to melt it in a tin on the stove first.

3 When the cast is hard, lift it up and gently pick and brush off any dirt sticking to the print. You should now have a beautiful positive impression of the footprint. It will take a few hours for the plaster to set as hard as it is going to, and so wrap it up to protect it on the way home.

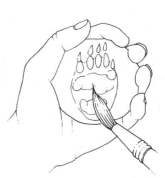

Step **3**

Plaster cast tips

- You can't take plaster or wax casts in the snow: the plaster or wax warms it enough to turn all but the hardest iciest print into slush before it has had a chance to set. In these conditions, photography is the answer. Just pay attention to the light source – the lower and further away it is from the vertical, the better it is at picking up the details of the impressions left in the snow.
- If you add too much water and the plaster is runny, mix in some salt to speed up the hardening process.
- Mix a little soap or PVA glue into the water and plaster to make the resulting cast less brittle.
- You can ink or paint the print to help it to show up. Alternatively, if you paint the surface with several layers of PVA glue or petroleum jelly and then press the cast into modelling clay, you can make a reverse negative cast of the positive you created in the field – you follow the same procedure, but the PVA or Vaseline keeps the new plaster from sticking to the old. Leave for a day to harden and then trim the edges of the cast with a knife until you see the line separating the two. Repeat all the way around and then gently lever the two apart with a knife blade.

Tracking beds

Away from the wet and soft stuff, the budding mammal detective has his or her work cut out. Tracking becomes a much more subtle art.

If you don't aspire to the sort of skills used by the rhino trackers in Namibia, you could try making a tracking bed. The idea is that when there is no soft substrate for a passing animal to leave its tracks in, you simply create your own. One use for this is to find out whether or not a burrow or hole is occupied. By smoothing the surface of the soil outside a Fox earth or Badger sett, which is usually fairly fine in texture anyway, you can, on return visits, tell when an animal has been in and out, and what it was. Then you can use a hand rake or a twiggy branch to wipe the prints out and start again. Obviously you should keep disturbance to a minimum, as it would defeat the object if your presence upset the animals whose front door you were investigating.

I've been using variations on this theme for years: I once tracked an escaped snake around my house using sieved icing sugar, I located the holes mice were using to access my shed and even found out what animals were passing through the hole in my hedge every night. Once in Uganda I discovered that a large aperture in a hotel's garden hedge had been created by a Hippopotamus that came every night to graze on the lawn outside my patio doors!

Wonderfully clear Jaguar tracks in soft wet mud on a river bank in Brazil. Prints are the only sign you are likely to get that this elusive nocturnal predator has been in the area.

I'd always thought these were smart bits of detective work on my part, but I have since seen the same idea used on a much larger scale as part of 'road ecology' studies in Canada. Tracking beds of fine soil about 1m × 50m (3ft × 160ft) were created across the width of an overpass as part of a study into how many animals of which species were using it to cross a busy stretch of highway. Daily checking revealed the tracks of nearly every mammal in the area from martens and squirrels to bears, Moose, Elks and wolves. A similar trick was being used down either side of a stretch of road to work out the ratios of animals of different species crossing the road successfully compared with those that were killed trying.

Tiger tracer

If time and storage space are issues or if you cannot be bothered to hump around the ingredients required to make a plaster cast, then a 'tiger tracer' is for you. This consists of a notebook-sized piece of Perspex on which you draw with a chinagraph pencil or non-permanent marker pen. You simply lay the Perspex over the footprint and draw around it, making notes of any points of interest. Later the print can be traced onto cellophane sheeting or paper to form a permanent, life-size record.

This technique got its name because it is used as part of Tiger studies in India, where researchers have found some Tiger tracks to be so distinctive in shape, size and other features

The keen amateur naturalist can use the tiger-tracing technique pioneered in India to keep records of footprints closer to home.

that individual animals can be identified by footprints alone. This is very useful for scientists trying to learn about the movements of secretive animals on a budget that prohibits the use of such invasive and expensive techniques as radio collars and GPS tracking. But the amateur naturalist can easily adapt it for use on a smaller scale and closer to home.

~ SPLITTING HAIRS ~

Whether it is a twizzle left on barbed wire or some fur found in a pellet or scat, hair has a lot to say. It's easy enough with the naked eye to tell the difference between the coarse, banded guard hairs of a Badger and the soft, fluffy hairs of a Rabbit, but under a microscope, the surface texture, colour, length and cross-section of each hair are very distinctive. This is usually seen as a specialist area of study, but if you want to take it further there are microscopic field guides, mostly aimed at a professional, academic market, that will tell you all you need to know.

The hairs that are of most use are the larger guard hairs. Under a microscope it is possible to see the cells and their arrangement in the pith, or medulla, of the hair. The scale pattern on the outside of the hair, the cuticula, is just as useful, but hard to see without a specialised staining technique. This obstacle can be got around if you take a gelatine impression of the hair and look at this instead.

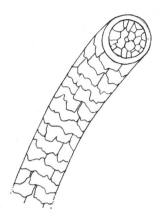

Deer hair and cross-section.

You will need:

- a microscope with 400× or greater magnification
- microscope slides
- a small quantity of gelatine dissolved in 10ml (2 teaspoonsfuls) hot water
- a pair of fine tweezers

1 Paint the gelatine solution onto a microscope slide, then leave it to cool and solidify for 10 minutes or so.

2 Using the tweezers, place the hairs across the gelatine film and leave them in position for 30 minutes as the gelatine sets.

3 Carefully peel the hairs up and off the film and look through the microscope at the impression they have left.

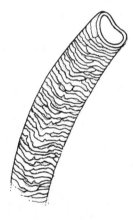

Field Vole hair and cross-section.

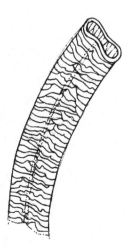

Rabbit hair and cross-section.

The rough stuff

the fascinating world of excrement

Scats, faeces, dung, spraints, whitewash, mutes – call them what you like, these are all terms used to name the waste products of digestion that have passed through an animal's gut and been excreted. Specific terms are often associated with certain species.

Droppings contain lots of information about diet and, in those species that use their droppings as 'signposts', about what the animal was up to: was it patrolling territorial boundaries during the breeding season? Was it passing through quickly and leaving a little and often? The study of dung is an eye-opener when you are dealing with the shy, elusive and nocturnal, and in many cases, it's as close as you are going to get.

I will never forget the look of shock on my mother's and brother's faces when they first found me face down on the front lawn poking around in a Fox's dropping with a stick. They simply did not understand, despite my very enthusiastic speech, that I now had a greater insight into the diet of our garden's regular nocturnal visitor and, what's more, I knew where all the raspberries were going!

Obviously the contents of droppings can give you clues as to what animal 'did it', but they vary so much in colour, texture and length depending on what the animal has been eating that this is not always enough. Size is an important guide, and the critical measurement is the width at the widest part. Specific animals have specific exit-hole dimensions! Identifying a small mammal species by its dropping alone is a little tricky. The droppings of insectivorous bats are superficially very similar to those of small rodents but are usually dry enough to crumble between the fingertips. Unlike those of rodents they have no real odour, either.

Bear droppings are usually large, blunt-ended, sausage-shaped masses often with evidence of indigestible items such as seeds seen here.

Location is often a big help, as is a combination of other signs – a large, pellet-shaped dropping combined with greasy smears on vertical surfaces next to runs suggests Brown Rat (*Rattus norvegicus*), while neat little piles of smaller pellets, near water and combined with runs and clipped vegetation are more likely to be those of a Water Vole (*Arvicola terrestris*).

Safety tip

There is a very good reason why humans are repelled by the odour of the excrement of many mammal species – survival. Excrement can harbour serious illness and all manner of bacteria and parasites, and so do not handle the droppings of any animal with bare hands if you can help it. Use rubber gloves or an inverted plastic bag to collect them and use a stick to explore them. (Remember which end you used, if you are going to pick it up again!) If you do come into contact with a dropping, wash your hands thoroughly with soap and water.

Bait marking

When scientist Hans Kruuk was working with Brown Hyenas (*Hyaena brunnea*) in Africa he noticed that these enterprising beasts had been scavenging the dead bodies of local tribespeople, who wore brightly coloured beads on their clothes and ornaments. The beads were turning up in the hyenas' droppings and accidentally yielding information on the territory boundaries and movements of the hyenas themselves. This observation was later artificially recreated and used in a technique that has since been called bait marking.

The sett is at the centre of a Badger clan's activities and so is a good place to start your mapping.

Bait marking works really well with a communally living beast such as the Badger (*Meles meles*). Begin by mapping your study area. Critical are locations of the burrows or setts, normally at the centre of a territory, and of latrines. In the world of Badgers, latrines are more than just holes dug for defecation – they are scented signposts with territorial significance, positioned at key points such as near setts, along territory boundaries, at places where the Badgers' well-worn paths cross natural boundaries such as hedges, and near any other important resource. A latrine can be anything from the odd single hole to a huge aggregation of pits on a boundary between two clans, where individuals from both sides make their presence known.

A Badger territory map showing the extent of two territories. You can make it as simple or ornate as you like.

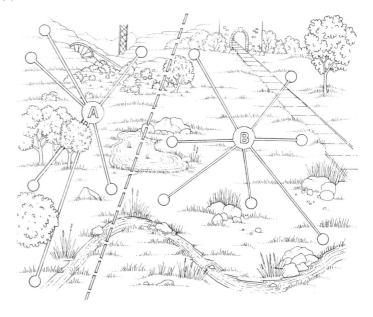

Once you have marked these key features on your map, you need a quantity 2mm (½in) marking pellets. Take a number of pellets of one colour, let's say red, and mix them up with unsalted peanuts and molasses or treacle. Then lug buckets of this heavy but Badger-irresistible fodder to what appears to be the major sett in the area. Deposit dollops of your goo around the sett, if possible hiding it from other creatures by covering it with leaves or a rock or log. (Badgers are strong animals who will not let such obstacles come between them and a good meal.) Now simply wait for a few days before starting regular tours of all the known Badger latrines to see where your red marking pellets turn up.

Every time they do, draw a straight red line on your map from the place where you left the food to the latrine. Over time you will get a star shape on your map which will represent the territory of the animals that live in the sett you marked with the red pellets. You can then repeat the experiment by leaving, say, blue pellets at a neighbouring sett. This will enable you to map out the next-door clan's territory and where red and blue pellets appear in close proximity, you will be able to identify the boundaries. It is an excellent way of getting a graphic image of Badger territorial life.

You can use a similar technique with small mammals, such as mice, substituting food dye for the pellets as they would give them indigestion. Small mammals tend to defecate when they are feeding, a 'bad' habit that the naturalist can tap into. Put your dyed bait (rolled oats are ideal) in a shallow tray and place this near a hole or a run.

A naturalist can learn about and map territories of Badgers by feeding them food and coloured pellets at the sett.

Then arrange other feeding trays with non-coloured bait in a grid-like system within the habitat. Cover each with a board or stone supported at the corners, leaving a big enough gap underneath for the animal to pass – this will help it feel comfortable while at the same time protecting your bait trays from wind, rain and thieving.

Check the trays after a day or so to see where the coloured droppings turn up, and you will begin to get an idea of territory size of the small mammals in your patch. Repeat with different colours in neighbouring habitats and territories, and plot your observations on a map as for large mammals.

Feeding signs

Evidence of eating is yet another sign to be interpreted by the naturalist in the field. It's like looking at a plate at the end of a meal. You would be able to tell whether the human had a T-bone or a take-out burger; well, the same goes for animals. Some are very tidy eaters and their feeding signs are hard to pick up; others would be easier to miss if the animal itself stood there with a big placard pointing the way!

Herbivores – nibblers and gnawers

By definition, many herbivorous animals spend a lot of their time taking chunks out of plants, whether chewing shoots and buds or stripping bark from saplings and tree trunks. They have individual table manners and have utensils and ways of wielding them.

Starting with the champion of chisellers, there is no creature quite like the North American Beaver (*Castor canadensis*). The teethmarks it leaves on twigs and branches are distinctive, having been made by a larger rodent than most, but it is the general modification of the environment that really gives the Beaver away: tree stumps and heavy-duty felling are about as extreme as feeding evidence can be!

The feeding evidence at the other end of the spectrum, though more widespread, takes rather more interpreting. Imagine trying to chew bark off a tree with your upper teeth missing. Why? Because that is what sheep and deer do – they have incisors only in the lower jaw and use these to remove bark by gouging at it. It is more of a scraping action than chewing or nibbling, and the result is fairly rough and ragged. (Incidentally, do not confuse deer feeding damage with deer 'fraying', an equally destructive act that leaves a tree denuded of its bark. With fraying, the motivation isn't a meal but either removal of 'velvet' from newly grown antlers or, in the case of a buck Roe Deer, territorial – leaving a visual and scent flag.)

The same job attempted by rabbits and hares produces a much smoother result as they tidily nip away at the bark, leaving distinctive-sized teethmarks with little or no fraying. Rabbits' and hares' teeth are so similar in size that they are hard to distinguish by measurement alone. But the hare's upper incisors have a notch halfway along their cutting edge, so each bite leaves two raised strips and looks as if it was made by an animal with four upper incisors rather than two.

Bank Voles (*Clethrionomys glareolus*) are very neat nibblers and will often nibble their way round a trunk at ground level (a process

The Beaver is the largest rodent in North America. A Beaver family can fell as many as 300 trees in a single winter, and a pair can gnaw through a 10cm (4in) thick branch in 15 minutes. Bark is the mainstay of their diet.

Guide to teethmark dimensions

Species	Distance across all incisors
Mice and voles	2.2mm (½in)
Water Voles and rats	3.5–4.5mm (⅐–⅙in)
Rabbits and hares	6–10mm (¼–⅕in)
Beavers and porcupines	10mm (⅖in)
Sheep and deer	10–30mm (⅖–1in)

known as ringbarking), killing the tree in the process. In a 'good' Vole year it can be hard to believe that the huge amounts of die-off seen in woodland are caused by these small rodents. They are also excellent climbers and will often sit in the 'v' of a branch and nibble away. This sort of damage almost glows white when it is fresh, though die-off at height is less common. Squirrel damage, on the other hand, is often obvious as die-off in the crowns or at the end of branches.

Look for the distinctive teethmarks of mice, voles and squirrels on fungi, too (slugs also rasp away at fungi, but they leave a depression and no teethmarks). Red Squirrels (*Sciurus vulgaris*) have a habit of taking their fungi and lodging it up high in the trees.

The same sort of logic can be applied to shoots and twigs that are chewed on by deer, rabbits and hares. Deer can bite only so far through a stem; the rest of the action is done by tearing, leaving a frayed end. Rabbits and hares, on the other hand, bite neatly through as if they were using a pair of quality secateurs. So now you can tell who's been eating your roses!

Similar observations can be applied to other browsers and grazers around the world, if you have some knowledge of their anatomy and feeding behaviour.

Next time you are out in an area where deer are found have a look for browse lines. This is the browse line of Roe Deer.

Little nutters

The smaller mammals from mice and voles to dormice and squirrels are real nut specialists, and their feeding remains are often encountered in the field. These become quite distinctive when you understand the mechanics of the 'chiggling' techniques used by each of these nutters to extract the nutritious kernel from the hard outer casing of a nut.

Squirrels take a notch out of one end of the nut and then, using their lower incisors like a crowbar, they split the nut into two halves. The halves will show the odd scratch made by the upper incisor.

Mice, with their long front limbs, hold the nut some way from their body and at an angle, with the top towards them. They nibble a hole and insert their lower incisors into it, using them to gnaw and the upper ones to grip the outside of the nut. Mice gnaw from the inside out, turning the nut as they go. This leaves a large hole at one end of the nut with distinctive incisor marks on the edge of the shell and a row of irregular scratches on the outer surface. The overall shape of the hole is often untidy, with the inner edge being lower than the outer.

Voles start in the same way, then insert their nose and upper incisors into the hole and use the lower incisors to gnaw away from the outside in. This leaves teethmarks only round the edges of the hole, which is a smoother shape than one chiggled by a mouse.

Dormice (*Muscardinus avellanarius*) do the tidiest job of all, leaving a very smooth lip to the edge of the nut, but a neat row of marks on the outside, as with mice.

Some of these small mammals also have a penchant for flesh – especially voles – and they use a similar technique. I often find caches of vole feeding remains containing nuts, seeds and a quantity of neatly chiggled snail shells! A rodent chewing on a snail shell tends to concentrate on the coils, unlike birds such as thrushes, which smash the shells into smithereens.

Voles and Rabbits commonly eat grass, too. Look for Field Vole runs under and around rough, tussocky grass – you will notice patches of neatly close-cropped grass, along with clippings and distinctive droppings. Water Voles leave similar traces on banks close to water. Rabbits are the famous creators of 'lawns' near their warrens which they habitually graze.

Grey Squirrel (*Sciurus carolinensis*).

Wood Mouse (*Apodemus sylvaticus*).

Bank Vole (*Myodes glareolus*).

Hazel Dormouse (*Muscardinus avellanarius*).

Pine, spruce and fir cones are another staple diet of many creatures. Their seeds are very rich in fats and oils, making them a particularly valuable resource in the build-up to winter. Climbers such as squirrels and voles will harvest them directly from the trees, while terrestrial species rely on those that fall to the ground or are missed at the dining table.

Squirrelled cones are the most obvious feeding signs when you walk through a coniferous wood. Squirrels hold the cone at an angle, with the base up, and systematically work their way

Several animals are able to prise the seeds out of pine cones for food. Squirrels are excellent at this, chewing the cone down to the core.

down the stem, removing the scales to get at the seeds. The fibrous remains of the scales give the leftover spindle a hairy appearance, with the last few scales often left undamaged; a clean cut through the stem is also usually evident, where the cone was bitten from the twig. You can even tell whether a squirrel is right- or left-handed by the way it has handled the cone!

A similar cone worked on by a mouse or a vole is much smoother, the base often nibbled into a rounded shape, and the evidence tucked away under cover, with the removed scales piled up near the scene of dining.

You might find the remains of cones that have been fed on by birds, too. Crossbills (*Loxia curvirostra*) do not remove the scales; they split each scale lengthways and twist and loosen the seed. The result is a tattered-looking cone, with the split scales being obvious trademarks. Woodpeckers just seem to batter the cone until it gives them what they want. A certain amount of splitting occurs here, but nothing as systematic as with the Crossbill's technique. The situation in which the cones are discovered also bears witness to the species concerned. Woodpeckers often jam cones into crevices in bark before attacking them, and only discard the old cone when they have a new one to work on – hence the cones and debris that may pile up beneath a popular 'forge'.

Often a tree becomes a favourite 'anvil' for an individual bird, because the depth and size of the fissures are perfect for the job. Such a tree often bears witness to this fact with cones jammed into the crevices.

Teeth

little miracles of design

Found in the jaws of mammals, reptiles and fish, teeth are wonderful devices whose form fits precisely with the function they evolved to perform. They help you to ascertain a lot about the owner, especially if they are still attached to its skull. Not only can you make a good guess at the identity of the animal by working out its diet and way of life, but you can also in some instances age a mammal by its dental records alone.

Incisors are the cutting teeth at the front of the jaw. The incisors of mammals such as rodents and lagomorphs (rabbits and hares) grow constantly through their lives and have to be kept in check by chewing and gnawing. In profile, incisors look like chisels. This is no accident, because the front face of the tooth is made of a harder material than the back, which means that the back and front wear at a different rate, resulting in a formidable set of self-sharpening blades that never go blunt. Rabbits and hares have a second set of tiny incisors behind the main cutting ones, a very good way of separating them from rodents by teeth alone.

Rabbit jaw showing prominent incisors.

Deer and sheep have no incisors in the upper jaw, just a bony plate, but the lower jaw shows four pairs, rather than the regular three (the fourth are really modified canines).

Bats have tiny, much reduced incisors. Not a lot of call for cutting teeth in these mammals, and they just get in the way of echolocating, and so they have been scaled down.

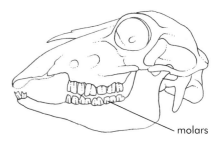

Sheep jaw, with no upper incisors.

Canines are the grasping, holding, ripping and gripping teeth. Also known as eye teeth, they are the four big, pointed teeth most visible in the jaws of carnivores. Certain deer and pig species and, of course, elephants, have these teeth enlarged for specialist feeding or combat. Rodents and lagomorphs have no canines and a distinctive gap in their place known as a diastema.

Molars and premolars are the cheek teeth, and their design depends on whether the diet is plant- or meat-based. In carnivores the cheek teeth are known as carnassials; they look like blades in profile and act like pairs of meat shears. In herbivores they are flattened to form grinding surfaces like millstones.

Dog jaw, with the sharp carnassials typical of carnivores, used for shearing through meat.

Long in the tooth? Estimating the age of an animal by its remains

It is possible to estimate a mammal's age by looking at tooth wear, as long as you have a rough idea of what a healthy, non-worn tooth looks like. Ten per cent wear for every year of a mammal's life is a rough average, so by the time it is five years old, its teeth will be worn down to about half their original length.

Age can be gauged more accurately if you have the chance to remove a tooth from a dead animal. Soak it in a weak solution of nitric acid (obtainable from good scientific suppliers) to dissolve all the calcium, then take a thin cross-section using a piece of specialist slicing equipment known as a microtome. When stained and looked at under a microscope, this section will reveal bands of growth around the root of the tooth. These bands are a material called cementum that holds mammal teeth into their sockets and is laid down annually in the autumn or winter. So the bands can be counted in much the same way as can the growth rings of trees.

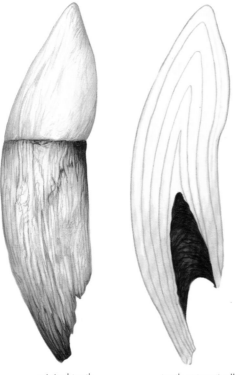

The tooth of a Humpback Whale cut to determine the age of the animal by revealing the number of annual layers in the core. The total number of layers can be seen only if the tooth is cut vertically. You can see the growth rings on the right, and the original tooth with root on the left

original tooth   tooth cut vertically

Bones – the hard bits

and what you can learn from them

Any vertebrate, be it mammal, bird, amphibian, reptile or fish, by definition has a skeleton inside holding it up and in shape, and forming a frame to which muscles and other organs are attached. Because the skeleton is so fundamental to a animal's form and to the way it works, even fragments can yield a lot of information.

The skull is a stunning lesson in form and function perfected. Just by glancing at one you should be able to say whether it belonged to a bird or a mammal. You can also judge the importance of the different senses. Large eye sockets obviously mean large eyes and good sight. The size of the ear bulla, though not quite as obvious, reveals the importance of hearing. Many channels for nerves and spongy bone hint at an acute sense of smell.

Mouse

Other clues about the owner's way of life can also be gained from the skull. In the case of mammals, large cheek bones and a sagittal crest (a raised ridge along the top of the skull) for the attachment of powerful jaw muscles say 'predator', a statement backed up by the nature of the teeth. The position of the eyes tells you how an animal perceives its world. Forward-facing eyes like ours give binocular vision, enabling us to judge distances – useful for leaping from branch to branch in a tree, for example – while eyes on the side of the head give better all-round vision.

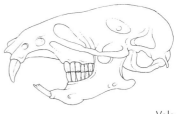

Vole

Tooth wear is the best way of telling an animal's age, but other clues can be had from the skeleton. The dimensions of the skull are a good start, especially if you have some known references as a comparison. Hares have a bony knob on the outside edge of the front leg bone (ulna) nearest the foot, which is very obvious until the bone finishes growing when the animal is about a year old. Bony growths and wear around joints (arthritis) can point to an older animal. You can also look at the jigsaw-like joins, called sutures, between the bones in the skull. These tend to start off very obvious and then fuse and disappear as the animal grows older.

Shrew

~ CLEANING BONES ~

If you are going to mount and display a skeleton, you'll want the bones to be as clean as possible. Nature has its ways of dealing with this process – there is a whole task force out there, honed to perfection by millions of years of evolution, whose sole purpose is to find dead animals and eat them. If you can get hold of a supply of these creatures and provide them with optimum conditions to get on with their work, they will eventually do the rest. This is, however, a smelly and time-consuming process: not everyone wants to keep a carcass crawling with maggots in the house, or even in the garden shed. Using chemicals is still pretty unpleasant – do it outside – but they do get the job done quickly.

You will need:

- a portable camping stove
- an old pan dedicated to bone cleaning
- a chemical solution of washing soda, caustic soda or sulphurated potash (you can obtain these from scientific suppliers)
- water
- a stick or pencil from which to suspend the bones
- string
- a de-greasing agent such as household ammonia solution, petrol or the sort of industrial degreaser you use on a car engine

1 Start with cold water and bring it to the boil *slowly*. If you put the carcass into boiling water you run the risk of toughening the flesh and making it even harder to remove. Tether the carcass or skull with string to a stick or an old pencil. It's much easier to pick up and check up on that way.

2 Add a solution of washing soda, caustic soda or sulphurated potash. These leave the ligaments in place, which means the bones stay in order. Note, the fresher the carcass, the better this works.

3 Check progress regularly, because if you leave them too long, or if the bleach solution is too strong, the bones become brittle and teeth have a tendency to fall out. (If this happens, you can glue them back in place with clear-setting modelling glue later.)

4 Once you have removed all the flesh from the bones, rinse them and immerse them in cold water containing a small amount of bleach for a few hours to sterilise and whiten them.

5 The last stage in the cleaning process is removing the grease. Using string, a mesh bag or a grille, suspend the bones in a de-greasing agent. The specimen has to be suspended rather than left sitting in the solution so that the grease drops off and sinks to the bottom.

6 Once you are satisfied that your bones are clean, rinse them thoroughly and leave them to dry.

*Safety note

It goes without saying that handling dead bodies has an element of risk associated with it. The body may harbour disease in the form of harmful bacteria or viruses. So precautions need to be taken. Avoid handling with bare hands – wear surgical rubber gloves. Don't get you face too close to your work and keep any open wounds well covered. Always wash your hands with a good anti-bacterial soap immediately after handling any animal material. Also take great care to not breathe in any dust from the animal material.

~ MOUNTING AND DISPLAYING BONES ~

Carry out the mounting process before the ligaments harden up, so that they are still pliable. In a perfect situation, if you have done your macerating correctly, the connective tissue will have kept most of the bones in their original positions. But it is more than likely that there will be a little fragmentation, and so you will have to use logic and imagination to decide where all the bits go and what shape the skeleton should rest in. A specialist anatomy book will help you, and with a little practice you will be able to recognise which bone is which and how they fit together.

You will need:

- wire of several thicknesses (depending on the size of the animal you are trying to mount)
- clear, quick-drying glue
- a polystyrene tile large enough to sit the whole skeleton on
- pins strong enough to hold the bones in place
- forceps

Step **1**

1 Start by threading the vertebrae and supporting them on an arch of wire, which will be permanently attached to the finished model.

2 Support this arch on a cradle of thicker wire and set the whole structure on the polystyrene tile.

3 Pin the limbs in position. Any missing joints or teeth can be recreated by the careful use of a dab of glue. Use forceps to lift small or delicate parts. It may help to use temporary supports to hold everything in position until the ligaments and/or glue have set firm. Then move the skeleton to its permanent display board and remove all the supports except that of the spine.

4 Attach the skull and support it on its own cradle. Label your specimen, stand back and admire.

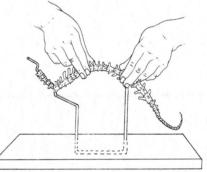

Step **2**

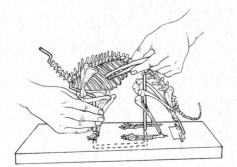

Step **3**

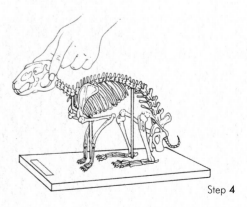

Step **4**

First trap your mouse

As we have seen throughout this chapter, you can learn a lot from the signs an animal leaves behind. But the smaller the mammal, the less likely it is to leave meaningful tracks, droppings or other evidence that is easy to find. Yet small mammals are out there and everywhere. Pretty much any ecosystem in the world supports a large number of them. In some parts of the UK populations of Common Shrews (*Sorex araneus*) are close to 50 per hectare (20 per acre). Because small mammals are important links in many food chains, studying them can shed light on the overall health of the environment on which they depend. But if they don't leave signs that you can interpret, you have to find other ways of getting close to them.

One common technique is the live small-mammal trap, which lures the subject into a trap and keeps it there until it is released. This is particularly useful for finding out what animals live where, how many there are and how they move round a habitat. Set your traps up in a grid, bait them without setting the trigger and wait a week to allow even the more timid individuals to familiarise themselves with them.

One of the smallest terrestrial mammals in the world, the Indian Pygmy White-toothed Shrew, having been caught in a pitfall trap (see p.105).

Once you have caught an animal, you need to mark it. Grip it firmly with gloved hands (even a mouse can turn into a miniature chainsaw when it thinks its life is at stake) and clip a patch of fur using fine scissors. This haircut doesn't have to be drastic, just enough to allow the individual to be identified a second time it is caught. Then release the animal and wait – with luck it will wander back into your trap again.

So how do you set about trapping? Well, what you call a small-mammal trap depends on what you call a small mammal – many of the same principles can be scaled up or down to suit. I have seen a trap design used to catch Wolverine work just as well with mice!

At the sophisticated end of the scale, the Longworth mammal trap is a well-engineered bit of kit that comprises two main sections: the trap mechanism itself and the nest chamber. An animal is lured into the trap by the placing of suitable bait. It hits a trip bar, which in turn trips the door mechanism. A gravity bar comes down and locks the door so that your mammal cannot escape and no others can enter.

Longworth traps are expensive and a bit fiddly to set up; they also come in only one size, which limits you to catching animals of about mouse size (though the sensitivity of the trip bar can be altered so that you can, for example, avoid catching lighter animals such as shrews). The one I pestered my mother to give me for my 11th birthday is still going, admittedly after a bit of riveting and repair, many years later.

The Longworth trap is made and widely used in the UK; the US equivalent is the Sherman trap, which is used worldwide and has the advantages of coming in a variety of sizes; of folding flat and opening out so that it is easier to clean, service and transport; and of having spare parts available via the internet.

If you're on a tight budget, plastic traps are available from pet shops. They are mostly designed for recapturing escaped pets and catching unwanted house guests in a humane way. They work, but small mammals can chew through some of them in a matter of minutes. If you want a cheap but effective trap, I recommend making your own.

The Longworth trap is composed of two main sections: the trap mechanism itself and the nest chamber. The animal is lured into the trap with suitable bait and hits a trip bar. This trips the door mechanism, bringing down the gravity bar on the door and locking it.

Getting your mammals out of the trap is the only tricky part of the process.

DIY traps take a bit more effort, but you can customise them to your exact requirements. A simple wooden box with a drop-down trapdoor held in position by a piece of wire bent around a nail and attached to the bait is all you need. When the animal takes the bait, it disturbs the wire, which then stops supporting the door; this swings shut and traps the animal. Make sure the door is a good fit, because if your enterprising captive can get its paws or teeth under the door, it can lift it and escape. Put in a couple of nail stops to prevent the door being pushed open.

Trapping tips

From the moment you set out to trap an animal, its life may depend on you – a responsibility not to be taken lightly.

- When baiting a trap, use a variety of food geared to the animal you want to attract. Raisins, seeds and unsalted nuts are obvious choices, but I have heard of people using bread soaked in aniseed, peanut butter, chocolate and cheese. Include insects, too, for the benefit of small insectivorous animals such as shrews.

- Remember that most small mammals are agoraphobes – they hate open spaces or stepping far from what is familiar. In the wild, set your traps against logs, tree trunks, roots, the edges of natural paths or anywhere under cover, among tussocky grass or a bramble patch.

- Be careful. I have unwittingly released many animals before I have had a chance to look at them properly, simply by being impatient. And be alert if your trap is surprisingly heavy – you may accidentally have caught a larger and fiercer animal than you expected.

- Do your homework first. Some small mammals are protected by law, and permits must be obtained in advance if you want to work with them. This applies to shrews in the UK, for example.

- Check your traps regularly, ideally twice a day, first thing in the morning and in the evening. Shrews do not do well in traps, and if you are likely to be catching them you should check every two hours or so. If you cannot find the time to do this, don't set traps.

~ PITFALL TRAPS FOR MAMMALS ~

Pitfall traps are easy to make and set. All you need is a smooth-sided vessel with a lid. Old biscuit tins, buckets and thick plastic storage jars are ideal.

1 First punch or drill holes in the bottom for drainage, then dig a hole the size of the container and bury it so that the lip is flush with the ground.

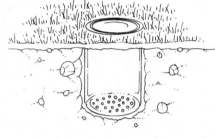

Step **1**

2 Trim the lid so that it just fits inside the opening of the container; then using glue or strong tape, make a pivot out of cocktail sticks or pieces of wire. When placed on top of the trap this makes a swinging lid that will tip any small creature that walks on it into the container below.

3 Cover loosely with soil, dead leaves and a large piece of wood or stone propped up at the corners. This makes the area attractive to small mammals, but more importantly it provides protection from the elements and prevents rainwater from entering the trap and making your captives cold and miserable.

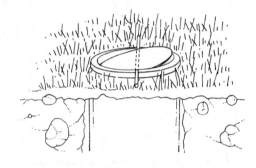

Step **2**

The one disadvantage of pitfalls is that you may catch more than one animal. This may not sound like a bad thing, but if you trap two rival males of the same species, or if the second animal is a predator such as a weasel, the outcome could be tragic.

Step **3**

Animals of the night

bats

Working with bats requires even more patience and specialist knowledge than working with terrestrial small mammals. The Chinese expression *yen yen*, which means 'swallow of the night', says it all to me – in their element, the air, bats are fast. They are also small, nocturnal, shy and protected by law. They squeak at an inaudible frequency and do not leave footprints. It's enough to make the naturalist give up on them and do something easy like go on safari.

But the fact that they are seemingly impossible to study makes them irresistible to the curious. Armed with the latest technologies, there is nothing to stop a 'Batman' or 'Batwoman' making some real discoveries – this is a frontier science.

Bats are highly successful animals. They have been around in a recognisable form for over 50 million years, and during that time they have diversified to fill numerous niches. There are somewhere between 850 and 950 different species – nearly a quarter of the world's mammals – ranging in size from certain flying foxes that have a near 2m (6ft 6in) wingspan and weigh 1.6kg (3½lb), to the world's smallest mammal, bumblebee or Kitti's Hog-nosed Bat (*Craseonycteris thonglongyai*), with a wingspan of 15–16cm (about 6in) and a weight of a mere 1.7g (¹⁄₁₆oz)!

Flying Foxes – an example of the Megachiroptera group of bats.

The family Chiroptera, to which all bats belong, is divided into two suborders: the Megachiroptera, a tropical supergroup of what we commonly call fruit bats, or flying foxes; and the highly diverse and wide-ranging Microchiroptera, which are found pretty much anywhere that there are insects for them to feed on. Here we will concern ourselves with the Microchiroptera, because unless you live in the Arctic or Antarctic, the chances are you have some living near you.

Most people's first experience of bats is seeing them... well not exactly seeing, but catching a quick movement out of the corner of the eye, a flicker by a light, maybe a fast zipping silhouette high above the head – a far from satisfactory experience for those wishing to know more.

My own introduction to bats involved a dead one – a Pipistrelle (*Pipistrellus pipistrellus*) that I found on my patio as a kid. I remember being fascinated by its minuscule size and the minutiae of its features – its claws, teeth and eyes. It was then dried and consigned to my childhood trophy collection, where it resided in a matchbox until the beetles found it. I didn't think much about bats after that until, during the twilight of one summer's evening, I noticed a few, probably Pipistrelles as well, patrolling up and down a line of oak trees outside my house. I don't know what prompted me to do this, but I remember thinking about how they located their prey with echolocation, and so I casually tossed a small stone in the air and watched in amazement as one of the bats suddenly changed course, ducking down toward the stone's trajectory... I had just interacted with an animal that was one of the biggest enigmas of the night, and that experience was quite enough to get me going 'bats'!

A Lesser Horseshoe Bat as few get to see them; not only are they very small, but they are very private and all roost sites are protected by strict laws.

Flying Northern Bat

The bat detector

Bats live in an acoustic world dominated by what is known as ultrasonic sound. It is very hard to describe – it's something like a sharp, high-frequency tick. Listen carefully next time you are out and, if bats are present, you may just pick up tiny bits of their calls. This is the lower range of the frequency of their communication; the rest we miss because our ears simply don't work at that level. Children's ears are more sensitive to these frequencies, and our ability to pick them up with the naked ear drops off quickly as we approach adulthood. But there is a readily available piece of technology that comes to the frustrated naturalist's aid: the bat detector, a simple hand-held device that acts as a translator.

Available from good natural-history and science suppliers, these boxes of 'electrickery' unlock the door on the mysterious world of nocturnal aeronauts. They are basically sensitive microphones that pick up the bats' ultrasonic vocalisations and then reduce the frequency to within a human's normal hearing range. You can listen to the sounds via a speaker or headphones and even record them for future reference.

Now do not get too carried away here. While it would be nice if each species had its own distinctive call, like birds, the bad news is that they don't. That would be far too easy! Bats produce a varied repertoire of sounds depending on what they are doing, their situation and the weather. On top of this, each bat detector is slightly different, and not all allow for direct comparisons of calls heard.

Some bats are easier to identify than others. The Greater and Lesser Horseshoe Bats (*Rhinolophus ferrumequinum* and *R. hipposideros*) make characteristic soft, warbling calls at two totally separate frequencies on the bat-detector dial. Noctule (*Nyctalus noctula*) and Leisler's (*N. leisleri*) Bats make a fairly distinctive 'metallic' noise. But the rest require a lot of walking about at night, good ears and intelligent detective work to unravel. If, however, you practise long and hard and hang out with other bat enthusiasts, you will soon get to grips with some of the more idiosyncratic sounds. A combination of the type of call and its frequency, the habitat and the behaviour of the mammals in question will get you well on your way.

If you want to encourage bats into your garden, the best way is to build a kind of bat box. This looks a bit like a bird nestbox, described on p.65.

Using a bat detector to count the number of bats leaving a roost in the roof cavity of a house.

Further afield

marine mammals

There is a whole group of mammal species which spend all or most of their lives in the world's oceans. This includes the pinnipeds (seals and sea lions), the cetaceans (whales, dolphins and porpoises), the sirenians (manatees and dugongs) and an otter or two. Studying these animals creates its own challenges – you don't find many footprints or feeding signs. With marine mammals, it's mostly down to direct observation, and this could mean as little as a far-distant 'blow' of water vapour or a glimpse of a flipper or tail. But even these can be tremendously exciting: while I admit that the sight of a vole nibbling on a nut may leave some people cold, nobody, surely, could be unmoved by a breaching Humpback!

The best way to see cetaceans is with an organised whale- or dolphin-watching group. Even if you normally prefer to do things on your own, a boat full of like-minded people squeezed together for the sole purpose of seeing these remarkable, smile-inducing creatures, added to the knowledge of the skipper and the usually present 'expert', is a recipe for a good time. As with a lot of mammal watching, you can be lucky or unlucky, but I am a great believer in the maxim that the more time you spend watching, the more you will see and understand. So let me tell you a story that will give you a taster of the rush you get from watching marine mammals.

It was going to take a lot of something to lift the spray-sodden spirits of the human contingent packed into this small boat. We had already been bumping and bouncing our way across the swell off the southern African coastline for an hour and a half, seemingly with nothing to look forward to but the same treatment on the way back – short of the boat actually sinking, things couldn't get much lower! But this is the way it is with cetacean-watching. Just when you think you have a round turn on things, nature throws something at you as a reward for your patience. In this case the reward was a Southern Right Whale (*Eubalaena glacialis australis*) and her calf, lolling around about 400m (440yd) away with the teal-coloured water breaking over the mother's distinctively encrusted, callused head.

A man watches Orcas from a coastal vantage point.

We cut the engines and the whale drifted closer to the boat, whose human contents were nearly falling over each other and into the water in the rush for the best observation spot. It was a 'Kodak moment' without a doubt, as the whale rolled over and we looked into her huge eye – not cold like that of a fish, but thoughtful – definitely that of a cognitive mammal. (If that sounds stupid, stare into any mammal's eyes next time you get the chance.) It was special, not because we hadn't seen this before – we had, maybe a hundred times in the last week. It was a combination of the intimacy, the build-up and the moment itself, very peaceful, with the silence punctuated only by the baritone exhalations of the cow and the more frequent, higher-pitched breathing of the calf.

As if to celebrate this, with all the pizzazz of a carnival, the scene was invaded by a different kind of breathing – little puffs, like air being released from a bicycle-tyre valve. It all happened so fast that it took a while to realise that we had been joined by a large school of Dusky Dolphins (*Lagenorhynchus obscurus*) and a couple of rare Heaviside's Dolphins (*Cephalorhynchus heavisidii*), speeding, leaping and weaving all over the place, in and out between the whales and our boat. Camera motor-drives were buzzing, although all I managed to get were a few perfect shots of dolphin splashes! It didn't matter – I will never forget the moment.

Watching and studying marine mammals can be a bit hit and miss – but when you get a sighting, all the waiting seems worthwhile. An added bonus is unlike with many other mammals – you're allowed to whoop with joy.

Right: Bottle-nosed Dolphins (*Tursiops truncatus*) in the Caribbean.

The earth creepers
Reptiles and amphibians

The herptiles, as they are sometimes known, are a group of animals that have a hard time in the world of classification. Although both groups are cold-blooded – more accurately poikilothermic, which means that their body temperature varies according to the temperature around them – reptiles and amphibians are not closely related.

The only obvious thing amphibians and reptiles have in common is that they nearly all (and even this is a generalisation) have a secretive and skulking habit. Even the word herptile, derived from the Greek for 'creeping thing', seems to back up our centuries-old loathing of creepy creatures. But combining the study of reptiles and amphibians under the heading of herpetology is quite handy, as many of the techniques used can be applied to both.

Right: A highly specialist snake, the Peringuey's Viper of the Namib Desert puts its simple body design to good use. Minimising contact with the hot shifting sand surface, it moves by side-winding or when the heat is on buries itself in the sand.

Left: Just mention their name and it conjures up lowly, slimy, creepy things. In fact the Amphibians and reptiles are some of the most spectacular, colourful and wonderful groups. The fact that much of their lives is a mystery makes them all the more appealing.

What's the difference?

The 9,850-odd species of reptiles and amphibians that crawl, slide, slither and hop over our planet's surface can be broken down into a number of smaller, more easily digestible groups. There are three orders of amphibian, numbering 4,550 or so species (see below). What makes these animals collectively amphibians is not easy to define and there are many exceptions to the rule, but most have moist, thin, non-scaly skin and a life cycle that at some point depends on water. The majority lay eggs which hatch into larvae that breathe through gills; as they metamorphose into adults, the larvae develop lungs – a neat trick that would be a useful way of defining them if it wasn't for the fact that some salamanders never actually 'grow up' and develop lungs. But, as a rule of thumb, if it hasn't got scales, but has four legs, a backbone and no fur, it's an amphibian.

There are about 6,000 species of reptiles and it is thought that they originated from an amphibian ancestor some 350 million years ago. The vast majority (5,700 species) are snakes, lizards and the worm-like lizards known as amphisbaenians; then there are the tortoises and turtles, the crocodilians, and two species of the lizard-like tuatara, found only in New Zealand. The main difference between reptiles and amphibians is that most reptiles are much less dependent on water. They have also evolved thicker, scaly, impermeable skin that helps them to conquer drier habitats, and either lay hard-shelled eggs or retain their eggs within the body, so are not tied to water for breeding.

Orders of Amphibians:

Anura (the frogs and toads, and some burrowing, worm-like animals)

Orders of Reptiles:

Squamata (the amphisbaenians)

Crocodylia (crocodilians)

Caudata (the salamanders)

Testudines (turtles and tortoises)

Rhynchocephalia (tuatara)

Gymnophiona (the caecilians)

The right handful

handling reptiles and amphibians

The golden rule is not to handle any animal unnecessarily, as you learn so much more if it is undisturbed, and in the case of herptiles, the word 'handling' implies any form of physical contact. However, there are times when it is essential if you are to identify an animal or catch it as part of a study. In these circumstances it is important to minimise the amount of discomfort and stress your subject suffers and, in the case of dangerous or venomous herptiles, to handle it as safely as possible.

Frogs are famous for their squirminess and slipperiness, and even a little one can be a real handful. Nets are ideal but still allow the frog to wriggle about or even jump out, making it difficult to establish specific details such as sex or identity. To keep hold of a frog – once you have finally *got* hold of one – cup your hands around it and then manipulate it so that the head faces out towards the gap created between the forefinger and thumb of one hand. As soon as it sees light and tries to break free, allow it to push most of its body through before clamping down gently but firmly on its hind legs – these are most frogs' 'thrusters', so if you can stop it kicking and at the same time support the front part of its body, you have it secure.

Frogs are slippery, slimy and strong. To hold on to them safely for inspection first make sure your hands are clean of soaps and insect repellent. Firmly but gently make a ring around the frog's waist with forefinger and thumb, gripping the legs with your closed palm.

Avoid dehydration

If you need to hold on to your specimen for more than a few minutes – and many species do take a while to identify, especially when you are working in the frog-diverse tropics – there is a risk that it may start to dehydrate from the warmth of your hands. So thoughtful herpetologists carry small, clear, freezer or zip-loc bags. Once your frog is inside the bag, you can view its features clearly and easily without either distressing it too much or inadvertently releasing it.

Toxic overload

You may think that amphibians are benign compared to famously dangerous herptiles such as snakes, but they have some surprising talents, especially when they perceive you as a life-threatening predator, and it is worth being aware of these.

Many species of toad and some frogs have noxious chemicals in glands in their skin; these are primarily designed to make them distasteful to a predator, although it has been suggested that the toxins also kill off bacteria and fungi that would presumably thrive in damp, warm conditions favoured by these animals. If ingested, some of these chemicals are very toxic to humans. So, as with all animals, wear gloves or wash your hands thoroughly after handling and do not rub your eyes, nose or mouth with your fingers. Take this warning seriously, as the secretions (bufotoxins) of some species, particularly the Cane Toad (*Bufo marinus*), have been known to kill humans. I have certainly heard stories from reliable sources of dogs and cats dying after just mouthing one of these animals.

parotid gland

Each bump on a toad's body is a poison gland. In bufonid toads, like this Cane Toad, they also have a huge parotid gland behind the eye which can eject a toxic fluid.

When stressed, the Cane Toad and other species can eject a toxic creamy white spray from the parotid glands on either side of the head behind the eyes. This can travel for up to a metre (over 3 ft) and once in the eyes, nose or mouth causes a severe burning similar to the sensation of eating raw hot chillies. I speak from experience on this one, as I was once retrieving my pet cane toad from a cosy little spot it had found beneath the television. It treated me as a predator, got me in the eyes – and there was no more watching TV that night!

I also have first-hand information which suggests that prolonged handling of some of the Poison-dart Frogs (*Dendrobates pumilio*) from South and Central America, even those species formerly thought of as fairly benign, can lead to some of the skin's toxic alkaloids being absorbed and causing nausea. In some species 0.00001g of these substances can kill a man, making the defences of these amphibians among the most potent in the world.

The Red-backed or Reticulated Poison Frog (*Dendrobates reticulatus*) from Peru.

Some of the large ambush specialists, such as African bull frogs of the genus *Pyxicephalus* and the horned frogs (*Ceratophrys* spp.) of South America, have a nasty nip on them, and the fang-like protrusions in the roof of the mouth can make their bites particularly painful. A little horned frog that I once captured bit on to the end of my finger and wouldn't release its grip – it even started swallowing and making its way up toward my knuckles, despite being only the size of a ping-pong ball!

Vivariums

studying and keeping reptiles and amphibians

Because of their small size and secretive nature, much of what is witnessed of the lives of reptiles and amphibians in the wild comes down to chance. Although you can substantially increase your luck by understanding your subjects' habits and being in the right place at the right time, making useful observations of anything from egg laying to mating displays is very hard in the field. The herpetologist can genuinely add to the world's knowledge by keeping and observing herptiles in captivity, and because of advances in science and a growing awareness of the problems facing these previously unloved denizens of our planet, this is an exciting and burgeoning discipline.

Much of what we know of these animals has been uncovered by herpetologists. Brood care in the Spine-headed Tree Frog (*Anotheca spinosa*) was systematically worked out in captivity, as was the fact that certain caecilians were loyal to the same tunnel system for long periods of time and not randomly pushing through the soil and leaf litter like earthworms. The most famous 'birth' in the amphibian world, that of baby frogs bursting out of the back of a female Surinam Toad (*Pipa pipa*), was first observed in an aquarium. All these observations can be applied to the wild, helping us to piece together an understanding of herptile ecology and making us better able to preserve these animals in the future. The other positive is that keeping and studying herptiles promotes an appreciation of a group of animals who, at the end of the day, need as many friends as they can get!

The female Surinam Toad carries her developing young beneath the skin on her back.

The very mention of the word captivity lets a lot of snakes out of the box, so to speak. It raises issues as to whether an amateur naturalist should keep wild animals at all, and brings to the fore concerns about the large international trade in herptiles for the pet industry, some of which involves very questionable practices. This is an area peppered with issues that are too complicated to have a place in this book. I will confine myself to giving some basic background information intended to start you off in the right direction. More specific information is available from books and websites dedicated to herpetoculture (see Further Reading p.336).

The basic piece of kit is called a vivarium – a broad term for any container or enclosure in which animals are kept for study. If you're keeping amphibians in a vivarium, it is essential that both your hands and the vivarium and its furniture are thoroughly sterilised and cleaned both before and after the release of your study subjects. Of course, the most important consideration when keeping any animal in captivity is what it requires to keep it healthy and happy. This is especially true of reptiles and amphibians, as they are highly sensitive to subtle environmental factors, so do your homework and prepare carefully.

Ultimately for an amateur naturalist, it's always best to observe various species under natural conditions in the wild. However, sometimes it might be necessary to keep some animals for a short time in captivity to study. If this is the case it's very important to be aware that some species in some countries are fully protected by law – such as Sand Lizards and Great Crested Newt in the UK – with this in mind where ever you are in the world, it's worth doing you homework and research first, if in doubt contact a local conservation organisation for help and guidance.

If any animal is taken into captivity – this especially applies to aquatic organisms – it's important that, they are kept for the minimal amount of time possible to allow your study and then put back exactly where you found them to minimise the spread of diseases, such as Ranavirus and Chytrid fungus, which are responsible for huge losses in wild amphibian populations. This applies as much to spawn and tadpoles as it does to adult amphibians.

The basic piece of kit is called a vivarium – a broad term for any container or enclosure in which animals are kept for study.

Frogging by night

The best time to see amphibians and reptiles in the wild depends, as with all creatures, on your position on the planet, the season and the weather conditions. But there are a few general tips to a happy herping trip.

My best herping moments have always been at night, from the place where it all started for me – watching Common Smooth Newts (*Triturus vulgaris*) doing their extraordinary aquatic courtship dance in my grandparents' pond and observing the incredible exodus of hundreds of tiny toadlets from the same pond – to the later experiences of finding the monstrous Goliath Frog (*Conraua goliath*) in equatorial Africa and Red-eyed Tree Frogs (*Agalychnis callidryas*) making their foam nests in the branches above a pool in Costa Rica.

The Red-eyed Tree Frog is nocturnal. During the day it sits on a green leaf, eyes shut and feet tucked in, blending with the vegetation.

The smooth, wet-skinned amphibians lose moisture readily, so it makes sense for them to be active primarily in the cool of the night. During daylight hours many hide under debris or in holes or, in the case of some tree frogs, attempt stunningly effective impressions of seemingly uninteresting foliage or bark. This means that your best bet is to go looking when they have relaxed a little and blow their own cover by going out foraging or searching for a mate (more about this later). I find head torches handy for this – they leave your hands free for scrambling and for using other tools and books. In the tropics, where many species are a long way above you in the foliage, I cannot recommend highly enough a halogen head torch. For me the extra range and brighter light far outweigh the disadvantages of having to replace batteries more frequently (I always use rechargeables and take spares with me just in case I end up being out all night).

If only they were all this big. A man in Equatorial Guinea gets to grips with the biggest frog in the world – the aptly named Goliath Frog, which has been reported to get to over 3kg (7lb) in weight and measure 32cm (13in) from nose to vent!

One of the reasons torches work so well is that much of these creatures' camouflage depends on light sources coming from the predictable direction of above; at night a torch beam catches them on the hop, so to speak. Also, many frogs and toads have a layer of reflective cells called a tapetum behind their retinas, so when their eyes catch your torchlight they reflect it back at you. This is especially effective when the torch is at the same level as your eyes, and reveals the position of many nocturnal animals, from spiders to elephants!

Third, and this is probably the most important reason, a torch beam focuses your attention. Being visually led primates, we are easily distracted during the day. At night our sense of sight is concentrated on the little pool of artificial light just in front of us.

Cracking the croak: the art of triangulation

My frustration at the lack of nocturnal frogging skills I displayed in the Venezuelan llanos will be etched on my mind forever, as will the simple solution. It was the second night I had spent up to my waist in murky, tepid water trying to find what sounded like a herd of chihuahua-sized frogs. I could not imagine how these animals could be so hard to locate when they were quacking and barking in the vegetation all around me, and by the sound of it, there were at least 50 of them scoffing at me. What made it worse was that, as I slowly waded in what I thought was the direction of my quarry, the little perishers would stop, only to start up somewhere else!

I had, it turned out in discussion the next day with a local herpetologist, made both an incorrect assumption and a foolish oversight. Firstly, just because a frog makes a big noise doesn't mean its body is proportionally large; and secondly, frogs are master ventriloquists. This makes sense if you are a male frog, as your calls advertise your presence not only to the females but also to potential predators (though how the females find the males is a mystery I haven't yet cracked).

Fortunately, my friendly herpetologist suggested I make herping after dark a social thing. Not only is it a good idea to have someone with you when you are wading around in piranha- and crocodile-populated swamps after dark, but there is also a literally 'sound' scientific reason for doing so. So that night there were two of us up to our thighs in the wettest swamp South America had to offer, and when the first frog croaked, I was introduced to the principle of triangulation. When you hear a call, both of you point in the direction you think it came from, then slowly move towards the spot where your bearings would cross if you were to draw a line. You keep moving closer, repeating the exercise every time your frog burps. It is a beautifully simple but effective way of working. By the way, the frog in question was only about 2cm (less than an inch) long and, other than the moment when its throat swelled up like a barrage balloon to utter its disproportionately loud bark, it did a perfect impression of a brown reed!

More often heard than seen, finding a singing frog is often harder than it might at first seem. It's worth persevering with in order to witness some fascinating behaviour.

Jelly babies

amphibian mating

The water bounces the blazing early morning March sky back at me; things are different in this weedy ditch this morning. The water seems to defy gravity and bulge upwards in the middle, with an odd lumpy texture. Every few seconds the whole reflection wobbles and shivers as something moves in the rushes at the edge, and straining my ears I can just about make out a bizarre purring noise, like a tiny electrical generator. The frogs are back.

Many years ago my 12-year-old heart would have jumped into my mouth on witnessing this scene, and I would have sprinted home to find a leaky bucket to pillage some spawn for a tank, the pond or even a bit of playground bartering. Nowadays not much has changed. I still feel that tingle of spring, but I know better than to interfere. My focus is held not so much by the inanimate spawn as by the activities of the frogs producing it.

Common Frogs (*Rana temporaria*) are the earliest of British amphibians to emerge from hibernation, occasionally breeding as early as January in the warmer southwest of England. Some spend the winter in the mud or debris at the bottom of the pond in which they will mate, taking in oxygen from the water through their skin, but most hibernate on land and, like all amphibians, head back to their breeding grounds as soon as they wake up. Being in situ when the females return may improve the males' chances of mating with them, but this is a risky strategy in a shallow pond, when they may be frozen solid if the winter is a harsh one.

This stage of the Frog's mating ritual is an unobtrusive one, mostly carried out under cover of darkness. If you stroll outside on a warm, wet night you are likely to spot Frogs around and in the pond, but not doing much more than bobbing about. Then at some unexplained but presumably hormonally triggered cue, the pond will erupt with Frogs splashing around in a frenzy of sexual activity. Mating will carry on 24 hours a day for the next one to five days, and in full swing is truly spectacular. Very little will distract a mating Frog from its purpose – even if one does notice your approach and either freeze or disappear

A frenetic frenzy of fornication: many species, such as these Common Frogs, spawn en masse.

Grab a partner

The mating hug, or amplexus, involves the male frog climbing aboard the female and hanging on with his front limbs. Rough, black swellings on his thumbs, known as nuptial pads, help him to grip. You can tell a male from a female because he is generally darker and smaller, with a bluish throat, while in some populations, the female is reddish. Once the female is ready to spawn (usually at night), the male will shed sperm on the eggs as soon as they leave her body.

It looks like the male Common Frog has a full nelson wrestling move on his mate, but this is a grip known as amplexus and many male frogs use it to hang on to their chosen 'love' and stake their claim to breeding rights.

below the surface for a moment, its surging hormones will soon convince it that you are no threat and it will carry on where it left off.

Even though you will probably be totally enthralled by watching this froggy orgy, you can still practise honing your naturalist's powers of observation.

Sometimes, when there are more males than females in the pond, you will see piles of males desperately trying to clutch on to a female but in fact clinging on to each other. In extreme cases, their frenzy may be such that they smother the female and she drowns. Even if a male manages to 'catch' a female, there is no guarantee that he will be able to mate – there will always be larger rivals eager to displace him.

The jelly that you see in ponds and ditches in the spring isn't all frog spawn. Common Toads (*Bufo bufo*) and our three native species of newt can be found breeding at the same time. Toads are fussier than Frogs, requiring deep water, and in any particular location generally emerge from hibernation up to a couple of weeks later. Because of this they breed in fewer ponds (one toad pond to every five used by Frogs). The mating activity of all the males in a colony tends to be triggered off at once, and their migrations are often large and visual affairs. Probably the best way to witness them is to contact your local Wildlife Trust and offer your services to their nocturnal Toad road-crossing scheme, designed to protect the sex-crazed animals from throwing themselves under the wheels of a passing vehicle in their desperation to get to their mating grounds.

Many frogs spawn en masse, here a pair embrace the wrong way around! As they struggle to get into position so the male can fertilise the eggs as the female lays them.

Newts turn up the colour and the contrast for their mating season. This male Alpine Newt is trying to impress a female with his finery and even a little dance.

Newts

Investigate a weedy pond in Europe at night, and by the light of your torch you may find any one of a variety of species of newt, Smooth (*Triturus vulgaris*), Palmate (*T. helveticus*) and the rare Great Crested (*T. cristatus*) among them. They return to water in the warm months of spring and, like their more vocal and better known relatives the frogs, they are there to mate and spawn.

Newt eggs are laid individually, often wrapped in the folds of a pond weed leaf, like this Italian Crested Newt egg.

What newts lack in noise, they make up for in poise, posture and colour. The males in particular are splendidly bedecked with membranous scalloped crests running the length of their body. With the coming of spring they turn up both the contrast and the colour; the flanks of all species become a rich collage of orange, blue, white and black. If you are lucky, you may witness a newt's courtship ritual in shallow water as the male fans his tail around the female's face, flamenco-style, and literally leads her on a merry dance. His choreography has to be precise as he deposits a capsule of sperm and directs her into position to pick it up.

Frog farm

growing your own

Watching an amorphous blob of proteinous jelly filled with black dots turn slowly from a collection of dividing cells through animated, wriggling shreds of life to free-swimming larvae and finally four-legged frogs is one of those user-friendly miracles that can occur right before your eyes given just a few square feet of space. But although there is barely a school nature table or a ten-year-old that isn't trying to witness this miracle, many of these well-meaning attempts fail. At certain stages in the life cycle some of the tadpoles fade and die for no apparent reason.

Well, having once been a frog-fanatical schoolboy myself, with many a failed froglet to my name, I have, via a process of trial, error and speaking to others with a similar interest, come up with the magic formula. Follow the steps below for a frog-friendly, foul-up-free and fascinating educational experience. This advice is based on the Common European Frog and the Common Toad, which make up the bulk of my personal experience, but the principles apply equally well to other species of *Bufo* and *Rana* found all over the world.

These eggs of the Red-eyed Leaf Frog are on the underside of a leaf overhanging water, safe from many predators. Soon they will wriggle free and drop with a plop into the pond.

Choose a 9–10 litre (approx. 2 gallon) plastic tank with a vented lid (those manu-factured by Hagen are excellent and very suitable for young and/or clumsy people). This a good size, neither too big nor too small and easy to move even when full of water.

If possible, use rainwater collected from a water butt. Failing this, any natural water from a pond or stream will do. If you have to use tap water, let it stand for a couple of days to allow the chlorine used to sterilise our drinking water to dissipate naturally. Many aquarium/pet stores sell dechlorinating water treatments, but check first whether these are suitable for amphibians.

Before collecting your spawn, have your tank set up and stabilised (see p.144 for setting up an aquarium), with a substrate of pre-washed gravel. Add a selection of pond weed to help oxygenate the water and stabilise its pH (acid/alkaline balance). The leaves will also soon become a meadow of microscopic algae plants – useful fodder for tadpoles. Try to use native weeds but, if you are collecting from the wild, screen the weed for 'predators' before introducing it to your tank – small specimens of predatory beetle larvae and dragonfly nymphs will soon become big, and with increased size comes a ferocious appetite for your tadpoles. A Great Diving Beetle larva (*Dytiscus latissimus*) will put away as many as ten tadpoles each day!

Resist the temptation to collect lots of spawn or tadpoles. Although you often come across huge quantities in the wild, only a few per cent of it will survive. So collect a small quantity of newly laid spawn – it should be quite firm and easy to separate with your finger. Half a cupful is an ideal quantity to achieve a ratio of three to five tadpoles for every litre of water (14–22 per gallon). Remember that the more tadpoles you have, the more work you will have to do. There will be more feeding, and more cleaning out to keep the water fresh, and if you have too many in too small a space they may slow down development and even turn cannibalistic. In nature, larger tadpoles in crowded conditions produce growth inhibitors in their droppings which, when eaten by other tadpoles, stunt their growth, a common cause of failure in rearing tadpoles in captivity.

Take spawn from garden ponds wherever possible – it keeps your impact and disturbance of wild populations to a minimum. It is also good practice not to risk contaminating a habitat by introducing spawn, pond weed or any other form of life that you have collected elsewhere. This is common-sense herpetological hygiene. Frogs in particular suffer from contagious diseases that may be spread unnecessarily in this way.

Collect your spawn with some pond water in a bag. To avoid temperature shock (remember those little black dots are living!), gently acclimatise your catch by suspending the bag at the surface for a couple of hours and gently mixing the warmer water of the tank with that in the bag, before finally tipping the spawn in.

One more tadpole down. The reason for so many eggs being laid is that many species find themselves on the menu. Here a Great Diving Beetle larva makes short work of a Common Frog tadpole.

Pond dippers at work.

Tank and tadpole management through the life stages

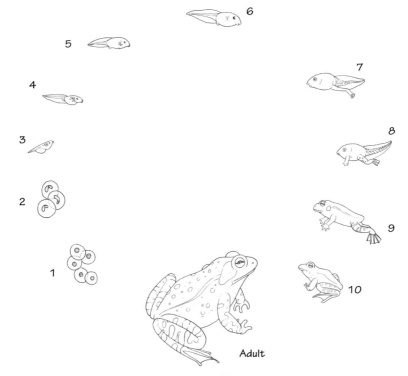

Adult

1. Freshly laid frog spawn.

2. For the first few days of their lives the tadpoles hardly move; they spend most of their time hanging from the sides and vegetation. They are not yet actively eating – they have no mouths – but are 'feeding' themselves by absorbing their yolk sacs. Do not disturb them.

3. The embryos hatch and become tadpoles. They will be content micro-grazing algae off the sides of the tank and any vegetation provided.

4. Once the tadpoles have started swimming around, change about half the water by tipping it out, using a net or sieve to catch any tadpoles that stray – they can simply be popped back in the tank. Replace the water with fresh, non-chlorinated water. Do this every week or so to keep a healthy-looking tank. If the water gets cloudy, change it more frequently. Remember not to subject the tadpoles to sudden temperature changes – keep the replacement water in the same room as the tank for a day to allow it to equilibrate. As well as providing a healthy environment for your tadpoles, you will promote strong and fast growth.

5. The head and tail are now more distinctive and the tadpoles begin to grow internal gill chambers for absorbing oxygen. The external feathery gills disappear.

6. After about three weeks they start needing more substantial salad. Every few days, give them a couple of the dry pellets usually intended for herbivorous pets such as rabbits. Observe how much they eat and try not to overfeed – otherwise you will have to increase the frequency of water changes or add other herbivores such as a few pond snails to do some cleaning for you. Introduce variety to your tadpoles' diet by boiling some lettuce for five minutes and suspending it a leaf at a time from the surface.

7. As the tadpoles grow, up their food allowance. Once the rear legs start to develop, they enter their carnivorous phase and suddenly develop voracious appetites. Feed them with flaked fish foods and small pond creatures such as bloodworms and *Daphnia,* also called water fleas.

8. As they start to develop front legs, your tadpoles' mouths and tongues grow and they will develop eyelids too.

9. As their tails begin to shrink, you need to reduce their food intake. They need a supply of very small hatchling crickets, aphids and other tiny terrestrial insects. At this time, you should also reduce the water level in the tank to just a few centimetres (about 2 inches) and provide plenty of haul-out space such as clean sponges, moss or stones.

10. Congratulations, you have froglets! Release them in the same pond (to avoid spreading disease) from which you collected the spawn. Do this after dark and in long vegetation, not back into the water.

Tails and scales

for the love of lizards

Lizards are fast. Well, that's the general rule – once warm and charged up by the sun, they become invincible, solar-powered rockets. This makes studying them quite difficult, and if you go for the desirable and uninvasive option of simply watching them, the only equipment that is going to be any use is a pair of close-focusing binoculars: the more distance you can put between yourself and your highly strung subject, the more likely it is that your presence will remain undetected and you will see natural behaviour rather than evasive action!

As a general rule, the bigger the lizard, the bigger its brain, confidence and ability to override natural caution when faced with humans in the wild. How tame and tolerant it is depends on such factors as whether or not it is hunted, persecuted or regularly fed or exposed to people. I have spent days trying to watch and get a good photograph of Water Monitor Lizards (*Varanus salvator*) in a national park in Borneo, where they were as elusive and wary as you might expect a big cat to be, never giving up much more than a length of tail or a blinking, suspicious eye. But at a resort just around the peninsula, the same species was being fed on buckets of kitchen scraps tipped into the Malaysian equivalent of a duck pond. The place was crawling with humungous, hand-tame lizards lolling about in the open. Some were even making bold approaches, too close for comfort in some cases – one of them lunged at my flip-flopped feet!

Female Sand Lizard sunbathing on a rock.

Solar energy

The best time to watch reptiles of any kind is when they are likely to be cold and trying to absorb solar energy. First thing in the morning is perfect, especially if the air temperature is relatively cool, the skies are clear and the sun's heat is just beginning to warm the ground. In some cases, this is the only time during the day that these animals will be out in the open away from cover.

An essential sunbath. It's how reptiles get up to speed.

Turn rustles into reptiles

Smaller lizard species are often overlooked simply because they are small. While the world is aware of the giant crocs and the monitor lizards, the really cool flying lizard (*Draco* spp.) or basilisk (*Basiliscus* spp.) is sitting close by, just as wild and beautiful but unnoticed. In fact many of the smaller reptiles catch your eye only when they skip off at your approach. If your senses are alert as you walk along a heathland path or a rainforest trail, you may well become aware of a rustle in the leaf litter a step or two ahead. Rather than ignoring it and striding on, it often pays to stop, make yourself comfortable and wait. Defocus your eyes, look in the direction of the noise and more often than not the perpetrator of the rustle – a lizard or even a small snake – will make a cautious return to its basking spot at the edge of the path, where if you remain stock still, you will be able to observe unnoticed.

Lasso me a lizard!

Whereas I really must stress that as with any living animal direct physical contact should be avoided unless strictly necessary, in some situations it is, however, a very handy way of obtaining information unavailable any other way. Just recently I was exploring among the aromatic creosote bushes and the rocky crevices of a gulley in the Sonoran Desert, in Arizona. This place is a herpetologist's dream; everywhere you look the place has a heartbeat, despite the seemingly impossible heat.

Now, stumble across a Rattlesnake or a Gila Monster and it is relatively easy to identify – as readily identifiable to the herpetologist as an Ostrich or a Kiwi would be to a birdwatcher – but the challenge comes with the reptile equivalent of the ornithologist's Little Brown Job or LBJ. All about small lizards skipped and scuttled between the boulders; with a little practice you could identify and separate different families but with some species, according to my field guide, I had to see the ridges running down their back, count their scales or even observe directly the colour of their belly scales! Tasks that are next to impossible with the use of eyes even with the aid of nice close-focus binoculars. For these you have to embrace another technique, known as noosing.

Flying dragon, its ribcage providing struts for its 'gliding' wings.

~ THE USE OF A NOOSE ~

The idea is simple – catching a lizard, but it is easy to spook a lizard with your shadow or by getting your noose snagged. A word of caution though: only use a noose on lizards (not on snakes or legless lizards) and only when you really need to. Some reptiles are fully protected by law, such as Sand Lizards and Great Crested Newts in the UK, so it is important to check the latest regulations before embarking upon such an activity.

To make a lizard lasso you will need:

The top end of an old fishing pole, or a hollow rod. You can get telescopic travel fishing poles which are quite useful too; and some kind of cord – nylon fishing line is fine but I prefer dental floss (and no it doesn't matter what flavour!); it is heavy enough not to blow around to much and holds its shape, which is handy when trying to slip it over the head of a lizard.

1 First you'll need to construct your noosing pole.

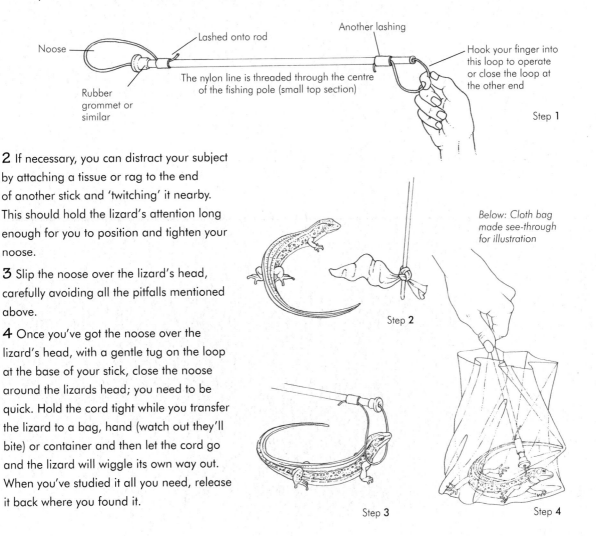

Another lashing

Noose

Lashed onto rod

Hook your finger into this loop to operate or close the loop at the other end

Rubber grommet or similar

The nylon line is threaded through the centre of the fishing pole (small top section)

Step 1

2 If necessary, you can distract your subject by attaching a tissue or rag to the end of another stick and 'twitching' it nearby. This should hold the lizard's attention long enough for you to position and tighten your noose.

3 Slip the noose over the lizard's head, carefully avoiding all the pitfalls mentioned above.

4 Once you've got the noose over the lizard's head, with a gentle tug on the loop at the base of your stick, close the noose around the lizards head; you need to be quick. Hold the cord tight while you transfer the lizard to a bag, hand (watch out they'll bite) or container and then let the cord go and the lizard will wiggle its own way out. When you've studied it all you need, release it back where you found it.

Below: Cloth bag made see-through for illustration

Step 2

Step 3

Step 4

Many lizards (and some snakes) engage in a somewhat distressing, self-mutilating survival strategy known as autotomy. When attacked, the lizard 'throws' its own tail, which will writhe in a rather gruesome way for some minutes, holding the predator's gaze long enough for the tail's former owner to slink off under cover and live to grow another. A lizard with a blunt, shiny-ended tail or different colours to portions of it has probably survived such an encounter. The reason for the different appearance is that when the tail regrows the bones inside do not, so once a tail has been lost once it cannot be shed again. When you are working with nervous lizards, autotomy can happen almost instantly with little provocation, but it is more likely to be avoided if you minimise stress to the animal, handle it as gently as possible and do not grasp the tail directly.

When alarmed or grasped by a predator or clumsy naturalist the tail of some species, in this case a Viviparous or Common Lizard, will come off and keep wriggling, a disturbing but effective survival strategy.

Tickling lizards: getting them down

This capture technique works particularly well with small tree-living lizards, especially those tricky, colour-changing, lightning-reflexed geckos that frequent hotel rooms throughout the tropics. It is also a handy little trick for performing the service that is often required of a naturalist – removing the said geckos from the apartments of less tolerant people! Often the act itself takes so long that you have plenty of time to persuade the occupants that leaving the animal in position will be better for all concerned. You can catch your lizard in a butterfly net, or similar.

Tickling requires nothing more than a long, thin piece of wire or even a length of robust grass or palm leaf and a small ball of kapok or cotton wool. Twizzle and tease the wool fibre around the tip of the 'tickling stick' so that it is firmly attached; then wiggle this conspicuous bit of fluff on the ceiling, trunk or wall your gecko is frequenting. Try to exploit natural cover, using corners and other obstacles to your advantage.

The skill in this bizarre puppet show is to make the fluff look like an insect target. You will find that your gecko's greed is much greater than any fears or doubts it may have. More often than not it will grasp the bait in its mouth and get its teeth temporarily tangled, so you can, if you are quick, bring it down or flick it into a hand net.

You can apply the same principle by tying a small insect such as a mealworm or even a recently dead fly to some thin fishing line. The lizards can literally be fished for – a great alternative to noosing!

Many species of lizard are adapted to live high in trees and on rock faces. Frustrating to get close to – but there are a few tricks that can be deployed to help.

'Road Riding'

There are limited ways in which automobiles can be used in a positive way – as opposed to causing undue disturbance – when watching wildlife, so you may be surprised to learn that they can be very useful when looking for reptiles, and snakes in particular, at night. In hot countries where most animal activity takes place after dark, our asphalt runways become a focal point. Why is a good question, and I have heard many theories. Some say that snake food such as mice and other small mammals occurs in higher concentrations here, perhaps because better marginal vegetation or edible scraps thrown out of cars attract it; others say the heat radiated back from the black surface of the highway raises a reptile's body temperature and therefore hunting ability. Whatever the reason, any self-respecting snake enthusiast should try road riding!

The best approach is to find a few quiet back roads that go through good habitat. Drive slowly along them a couple of hours after dusk, using the light of the headlamps to watch for animals at the side of the road or on the road itself. Keep a head torch on your brow and your handling tools (see p.133) at the ready, because you do not know what is literally around the next corner.

Obviously if the roads you have chosen are public highways, you are going to have to make some compromises. Just because you have seen a coral snake on the verge and jammed on your anchors, do not assume the car behind you will do the same!

A safe alternative and something I have done in Tucson, Arizona, is to befriend the local golf-course owner and persuade him to let you borrow a golf cart. The roads on golf courses are fantastic places to pick up nocturnal animals, as by nature they contain or adjoin semi-wild habitat and are often a source of unseasonal water and productivity. Here you can safely and at your own speed go whizzing about to your heart's content.

Roads and tracks are a good place to spot reptiles and amphibians at night. Partly they show up better in the open, but also some are attracted to the warmth radiating from the surface.

Golden rules

I have heard of some horrific near-death experiences caused by the sort of behaviour described above, so here are a few more golden rules:

- Always be aware of other road users.
- Give way to other cars.
- Do not make sudden manoeuvres.
- Always pull over in a safe place, avoid stopping on corners or bends and avoid busy times.
- Use your blinkers when stopped.
- Don't point torches directly at other drivers.

Snakes that pack a punch

All snakes, from an angry cobra to a Grass Snake you have just hooked out of the pond, are capable of inflicting a bite that will at best be painful, and at worst can lead to agony and even death. I tend to treat all wild snakes in the same way – with oodles of respect and caution. A friend and experienced herpetologist who was bitten by his pet python had fragments of the snake's teeth stuck in his hand for months afterwards and the wound was slow to heal due to infection. Even small, seemingly inoffensive species such as rat snakes (*Elaphe* spp.) could have been eating rodents that themselves were carrying a nasty disease. So the way you secure, hold and support a snake is pretty much the same whether it is venomous or non-venomous.

There is a worrying trend in the popular media to show and glamorise the activities of 'reptile wrestlers' and 'fang fiends'. These people are certainly very knowledgeable and skilled at what they do – well, obviously; if they weren't they'd be dead! If you are tempted to emulate these iconic figures, remember that you may be putting yourself, and more importantly the animals, at risk. Having said that, my stance is to give as much responsible information as possible and let people make up their own minds – after all, it was early exciting experiences with Adders and Grass Snakes that led to my life long fascination with snakes in general.

Indian Cobra (*Naja naja*). Carefully does it... never handle reptiles unless you are certain you know what you are doing! A mistake could be your last.

Notes on handling the small and the venomous

Nearly all snakebite-caused fatalities are the result of people trying to handle venomous snakes and underestimating the animal's capabilities. No-one, whether amateur or professional herpetologist, should ever handle a venomous snake without the correct equipment– anyone who does needs their head testing. The best herpetologists are the ones who live longest and get to shout out loud to the world what fantastic animals herptiles are. To prolong your life expectancy, it is wise to follow some golden rules.

Never, ever touch a venomous snake. Even recently dead ones need to be treated with caution.

With large and strong species, be aware that the

A species not for the amateur; fortunately most venomous species are not quite so 'in your face'. Even this spitting cobra would rather flee than face you.

animal may try to loop its body around itself and form a slip knot which it will use to push your hand off its head. Pay attention and be ready to twist away before this becomes a problem.

Avoid complacency at all costs. Do not lose concentration or ease up your firm pressure, as however subdued a snake may seem it can suddenly spring into action and catch an unwary handler out.

For normal study there is no need to touch a snake's body. You can purchase numerous tools from specialist dealers on the internet or simply manufacture them at home: these include all manner of hooks, grips and tongs and are designed to maximise the safety of both animal and handler. I really recommend using these, although they can be expensive.

I recently spent some time with the Irula tribe in India who, for the purpose of venom extraction, handled kraits, cobras and saw-scaled vipers on a daily basis. These masters of snake work used nothing

Hooks and snake sticks

Hooks and snake sticks tend to be more gentle and sensitive if used correctly and are among the most useful and wonderfully simple tools a herpetologist can have. But in the hands of an excitable and inexperienced person they may grip the snake too tightly and cause internal damage, breaking ribs and damaging nerves. To reduce this risk, glue foam-rubber pads to the inside of the jaws of the tongs, or purchase one of the more expensive 'gentle' types. Whenever possible, grasp the snake halfway down the length of its body and never grab it behind the head or even support it like this for the briefest time, as you may kill it.

but their bare hands and an iron pole to dig the animals out of their holes. Working alongside them with all my specialised tongs and hooks made me feel a bit stupid, but having said that it is better to stick with what you are used to.

Hooks and sticks come in a variety of sizes from telescopic, pen-sized ones (which frankly are of limited use, but are handy for poking around in dark crevices if you are looking for scorpions and spiders to huge unwieldy poles that need two hands to operate them and are designed for use with large pythons and constrictors in captivity. Hooks with a longer 'flat' at 90 degrees to the handle are the most useful for pinning and tailing. If you are going to be lifting snakes up, you need a hook with a 'neck' that bends away from the main shaft at approximately a 45-degree angle.

Commercial snake sticks consist of a length of tempered steel with a golfing grip at the other end; and for the travelling naturalist collapsible telescopic ones are very useful. The type and size you need depends on what you are going to use it for and what species you are likely to encounter. The stick should be at least 30cm (12in) longer than the longest strike range of the snake you are dealing with (that is at least half the body length for most species, although some can launch themselves almost the entire length of their body, so as always expect the worse and be super cautious). Longer sticks can be awkward to manoeuvre, especially in the close confines of vegetation. Using two at the same time, like a kind of simplified Edward Scissorhands, is often the way round this problem, and with faster, non-venomous species a hook can also be used in conjunction with the 'tailing' technique shown below.

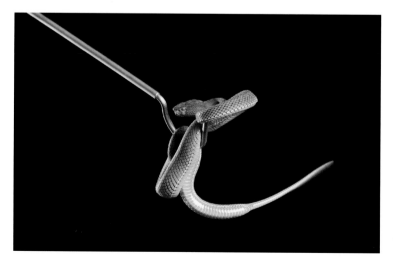

Lighter snakes such as this arboreal green viper can be lifted with one well-placed hook – just be careful it doesn't climb up the handle – a good reason for using a good length one.

~ MAKING YOUR OWN SNAKE STICK ~

If you are on a budget, you can create a very workable and usable set of snake hooks by bending and modifying various diameters of welding rods, or even a section of a thick wire coat-hanger. Note, however, that handmade snake sticks are too flimsy for pinning or lifting heavy species. For such purposes use a thick gauge wire instead.

You will need:

- a suitable piece of metal
- a length of wooden dowel

1 Hammer the tip of your metal flat and round it to make it easier to slide under the animal's body.

2 For smaller hooks use a bit of coat-hanger (one of the straight bits, not the hook itself) and secure it firmly to a length of wooden dowel.

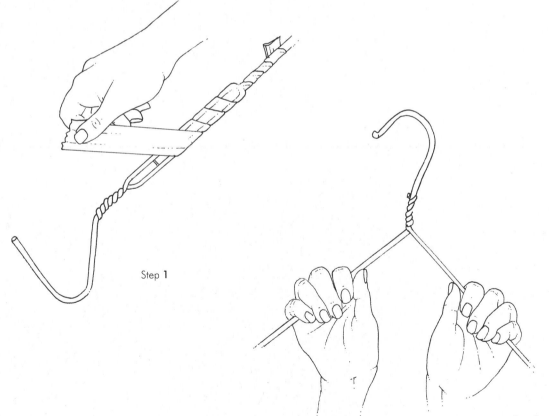

Step **1**

Step **2**

Using a snake hook or stick

There is no substitute for experience when it comes to using a snake hook, so practise on the smaller, slower, non-venomous species. For terrestrial species, those that live most of their lives with their bellies hugging the ground, slide the hook under the snake's body at the halfway point and gently lift it up. This is usually all you need to do, as most snakes have a reflex response to height and the possibility of falling; they will freeze and grip the hook as best they can. As long as you keep the hook high enough for the snake to feel it cannot crawl down, it will stay put. Be careful not to lift the snake higher than your gripping hand, as this encourages it to move towards you; and be aware that although the snake isn't going anywhere, it may well feel threatened and strike out – keep a sphere of safety around the snake and be aware of the position of any other people, too. Heavier snakes, particularly Gabon Vipers (*Bitis gabonica*), Puff Adders (*B. arietans*) and other big-bodied vipers, can be injured by their own weight pressing on the hook; supporting them with two hooks distributes the weight better.

A researcher uses a snake hook to catch a Forest Cobra in Cameroon, carefully keeping his distance.

For faster colubrids and other non-venomous snakes, using a snake stick is an art form akin to juggling spaghetti with chopsticks! At the same time as you pin the snake with the stick, you want to make a confident grab at the tail. This is known as 'tailing' and it relies on holding the snake by the rear end of its body while supporting and/or controlling the front end, which in most cases will be either trying to get away at all costs or coming back at you. Those that crawl up the sticks (particularly arboreal species) or try to get away are best dealt with by using two sticks. These 'runny' snakes usually settle down after they have passed back and forth several times; those that don't need to be restrained by being 'tailed'. Pinning sticks are something you shouldn't need to use. Pinning is a way of restraining an animal so that it can be picked up while gentle pressure is exerted on its jaws to prevent it from biting. This technique is mostly used in laboratories where snakes are manipulated in order to extract venom. For the amateur, it is just asking to be bitten.

A quick word on transporting snakes

For everything from big anacondas to the tiniest of colubrids, the best way of transporting a snake is in a cloth bag of an appropriate size – those cloth laundry bags available from the top drawer in posh hotels are great because they come with a built-in drawstring, but a pillow case or cloth sack will do fine. Place or drop the snake into the bag, use your hook to squeeze it gently towards the closed end, twist the neck of the bag together and tie it back on itself. The snake is now safely trapped in the dark, soft confines of the bag. Do not be complacent, though: just because the snake is out of your sight doesn't mean that you are out of its sight or bite! It may be able to see you through the weave. Long-fanged species especially can and will strike out at shadows, and being bitten through a bag is no laughing matter. Cloth bags don't give their occupants much physical protection, so place the bag in a bucket, tub or more robust bag while the snake is in transit.

Showing a snake a dark bolt hole, in the form of a pipe leading to a bag, is one of the safest ways to persuade them to go in and then transport them.

Rules of engagement

- If you have doubts about the identity of the animal or your ability to handle it, leave well alone.
- Never handle large or venomous reptiles if under the influence of intoxicating substances. Always have your wits about you.
- Never directly handle or pin a venomous snake.
- Know your limits and do not be afraid of admitting them – when it comes to safety, leave your ego at home.
- Never work with potentially dangerous reptiles alone. Always 'buddy' up with someone reliable.
- Be aware of the current laws before engaging with any reptile for the country you are in, see codes of conduct p.11.
- Don't catch reptiles unless you need to move, study or identify them.

Fish fantastic

It's a shame that, unless they are dangling off the end of a baited line or nestling on a bed of greasy chips, most of us tend to ignore fish. We rarely make the effort to get to know them as we do other, more obviously charismatic life forms. But there are ways in which we can enter their watery realm, if not exactly on equal terms, at least in such a way as to begin to understand a little about what makes them tick.

One thing that leaves fish swimming against the current of interest is that they are often regarded as stupid and uninteresting. When faced with this in conversation, I tend casually to bring up swimming with a Great White Shark (*Carcharodon carcharias*) in the cool teal waters of South Africa or sharing space with the second largest fish in the world, the rare and mysterious Basking Shark (*Cetorhinus maximus*). For me, these two experiences knock Polar Bears and Mountain Gorillas into a hat – they're the biggest adrenaline trips I have ever had and both are undeniably fishy ones! Whether we are talking toothy top predators or the smallest of small fry, fish are quite simply fascinating.

Below: Barracuda shoal.

Left: Shark skin detail.

Fish, fishing and figuring it out

My introduction to the world of fish came from fishing trips with my father. It dawned on me very quickly that to be a successful angler I had to understand not only where the fish were but why they were there. Even that wasn't enough for me; I wanted to learn about the fish themselves. I wanted to see them doing what my father described to me. I longed to see the Perch (*Perca fluviatilis*) hanging under the moored boat, sun dappling their tiger-striped backs, maintaining their position by the lazy flicking of red fins. I needed to see a Trout (*Salmo trutta*) waiting patiently, like a Leopard, in the shadows by the eddy, picking off the occasional stray Minnow (*Phoxinus phoxinus*), and of course I hungered to see the Minnows themselves. Catching the fish, ripping them out of their world, lifting them through their sky into ours, soon started to lose its magic, just as the fish, flipping around in my hands, gasping and asphyxiating in the air, lost theirs like a wet pebble drying.

The deciding moment came one day when I was staring at my float, watching it bob and swerve on the surface, knowing that every twitch and ripple had an equivalent action, caused by the fish just below the surface. If I stretched out my fingers I could touch it, but it remained invisible, cloaked from my view by the turbidity of the water and the play of light on the surface. It was both frustrating and tantalising for a young boy. Wouldn't it be great, I thought, to be able to relate what my float was doing to the behaviour of my quarry under the surface? The question in my mind was how.

It was then that I stopped catching and eating fish and started keeping them.

For most of us our impressions of fish are usually confined to what we catch with line or net, or what we see on the fishmonger's slab.

All tanked up

As a naturalist, you will quickly find that you become a connoisseur of containers, vessels and canisters. Few containers are as useful as a transparent tank. Home-made or shop-bought, they are very handy and versatile things. Fill them with water and they are aquariums, keep them dry and they become vivariums for terrestrial animals. When it comes to materials, the most common question is 'plastic or glass?' Both have their advantages and disadvantages, as you can see in the table below.

Choosing plastic or glass

	Glass	Plastic
Cost	Tends to be more expensive.	Cheaper if bought new.
Weight	Heavy and awkward.	Light and portable.
Clarity	Tends to stay looking good for longer and is harder to scratch.	Can become opaque with time, and is a nightmare to clean without scratching. Even if you are very, very careful, this will happen.
Repairs	A leaky aquarium can be repaired more often than not with clear silicon sealant but attempts at repair are usually futile!	Forget it. Slightly cracked ones that once contained water can be reincarnated as dry vivariums.
Fragility	Chips easily and can crack if placed on an uneven surface or exposed to temperature extremes. Watch out especially when cleaning with hot water – never make it too hot.	Tends to get brittle with age but is fairly robust. Some types of plastic will melt if you put heat mats under them.

~ SETTING UP AN AQUARIUM ~

There are four basic types of aquarium: saltwater, freshwater, tropical and cold. Freshwater is the easiest to set up, and I suggest you stick to this until you are confident of what you are doing. There are loads of different kinds of tank available in many different sizes. Choose the biggest you can easily maintain. You can have too small a tank, but rarely one that is too big for its inhabitants. There isn't room here for instructions on keeping individual species of fish, so for advice on that check the back of this book for further reading.

You will need:
- a tank with a close-fitting lid
- clean natural water (see 'A word on water', p.145)
- an old toothbrush
- enough pea gravel to cover the bottom of the tank with a layer 2–3cm (about 1in) deep
- a bucket to wash the gravel in
- 'furniture'(rocks and/or wood)
- enough cork tiles to form a base for your tank
- old paper, card or a plastic bag
- filter (optional)
- pond weed and inhabitants

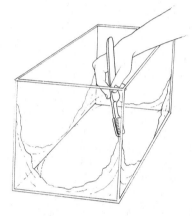

Step 1

1 New or old, give your tank a good wash – you never know what contaminants, dust, vapour or disease it may have been in contact with and it is best not to take any chances. Use an old toothbrush to clean out the awkward corners.

2 Wash and scrub the pea gravel by placing it in a bucket of clean water and swilling, scrunching and rubbing it. Discard the water and repeat the process until the water runs clean. Clean and sterilise any other equipment or 'furniture' you may be placing in the tank. There are various chemical sterilisers on the market that are harmless to pond life, but I find using boiling water straight from the kettle effective enough.

3 Place the tank where you intend it to stay, as a tank full of rocks, water and living things is heavy, awkward and fragile; moving it will also cause unnecessary stress to those ensconced. If you are not using an artificial light source, choose a light position out of direct sunlight, as temperatures can soar and spell certain death to your fish; algae can also become a problem. Place the tank on a level surface. I find cork tiles (available from DIY stores and pet shops) very

Step 2

useful here – they act as a cushion, levelling the bottom of the tank and minimising stress on the glass that can cause it to fracture when full. (Believe me, I talk from experience when I say it is no fun to have even a little tank burst in the middle of the night and to find yourself tiptoeing around in bare feet on a soggy carpet, trying to avoid shards of glass and rescue the unhappy pond life that has been unceremoniously dumped on the shag pile.)

4 If you are using an under-gravel filter, place the tray and associated gubbins into position as per the instructions that came with it. Otherwise gently scoop the gravel into the tank, landscaping it so that it slopes down towards the front. This makes it easier to clean and simply looks better.

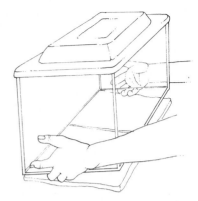

Step **3**

5 Arrange the furniture, filters and other equipment, but do not plug in or turn on yet. Place the paper, card or plastic bag inside the tank and pour water onto it; this way the turbulence created doesn't disturb the gravel or any leftover silt or debris that might cloud the water. Don't fill the tank to the brim – leave a little space at the top to help prevent things jumping out. Remove the lining material.

6 Now add the plants. These take up waste nutrients produced by other life in the water and recycle it; they also help aerate the water by producing oxygen – plus they look fantastic and provide good shelter and living furniture for the creatures you are planning to keep.

Step **4**

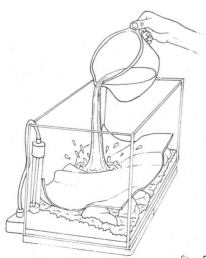

Step **5**

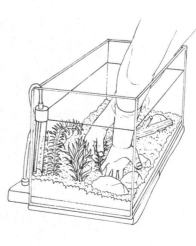

Step **6**

A box of water – aquariums

I began taking some of the fish I caught home to study. Keeping them in the most natural conditions I could create, I was able to find out about living fish in their own environment. To start with I even fished in my own aquarium – without hooks, I hasten to add – just so that I could relate the movements of the fish to those of the float. But soon the fish themselves took the lead. Their movements and their jewel-like quality in the clear water became intoxicating and, like anyone who has ever stared into the fish tank in the dentist's waiting-room, I found myself, well, hooked.

Top tank tips

Put on a tank top. Many pond insects can fly – that's how they colonise new watery habitats – and some fish and frogs can accidentally flip themselves out of the water or simply do a Houdini as part of their natural explorations. Make sure you fit a good lid and that escape is impossible. As I write, a snapping turtle in a tank next to me has just lifted a heavy metal and glass lid weighed down with bricks!

Make sure your tank is out of reach of troublesome dogs, cats and kids. I have had curious cats spend the afternoon systematically fishing out every unfortunate tadpole, leaving them lined up on a windowsill, a pathetic little death row of dried bodies. Also, a large and thirsty basset hound once drank a whole tank of water along with

most of the occupants. Both these events could have been avoided by positioning the tank more carefully, and they illustrate the importance of a good, tight-fitting lid.

A word on water. Natural rainwater is always best – collect it in a water butt. If you have to use tap water, put it in the tank and let it settle for a week or so before adding any animals. Alternatively, use one of the many water conditioners and dechlorinating solutions available from specialist aquarium suppliers, including most pet shops, and follow the manufacturer's instructions.

Electricity and water do not mix. Always turn off electrical elements such as pumps, filters, aerators and water heaters before placing your hands in the water. That way you do not risk electrocution. Putting a circuit breaker on the mains supply is an added precaution. Also try to avoid the mains power supply being directly below the tank.

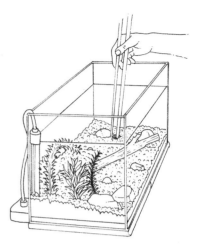

long chopsticks

Use long chopsticks or make some long tweezers out of split garden canes to move things around with minimum disturbance. Turkey basters are useful for sucking up small, offending articles from an established tank.

Do the housework. Look at your tank as a kind of aquatic window box. Just as you need to weed your pansies and petunias, your tank will also need a little gardening from time to time. It is perfectly natural for debris to accumulate on the floor of the tank, but it can be kept under control by regular hoovering. The easiest way to do this is to siphon it up through a length of clear rubber tubing. Place one end in the water with the other end at a lower level, outside the tank. Give a short, sharp suck on this end until you see the water coming down the tube, then very quickly take the siphon out of your mouth and direct it into a bucket or similar receptacle. Now you can direct the other end around the tank, sucking up sediment. You may get an occasional mouthful of water to start off with but, since this water can taste pretty bad, you soon learn to avoid it! On a serious note, though, certain aquatic reptiles and other animals carry germs which at best can leave a foul taste in your mouth and at worse make you very sick. Cheats can spend top dollar on a purpose-made siphon, which has a flexible bulb on the end and alleviates the need to suck. Or you could modify the bulb off a turkey baster.

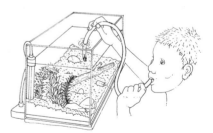

siphon tubing

Get in the flow by recreating conditions in the wild. Bear in mind that many species live in flowing water and will die of suffocation in a matter of hours in still water. To overcome this you *must* provide a pump and aeration. If this is too specialised for you or you simply cannot afford it, stick to still-water species.

~ THE 'JILLY' JAR ~

I have absolutely no idea where the name of this fish-catching technique came from – all I know is that a bunch of my school friends and I would get together every Saturday morning to go 'jilly-jarring'. Maybe we invented the term? Maybe it has some colloquial origin deep in the backwoods of East Sussex? Who knows and, as long as it works, who cares? The size of the jar doesn't really matter – a bigger jar doesn't necessarily catch more fish, and the extra weight of water the string has to support when you pull it out may be a problem.

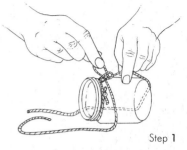

Step **1**

You will need:

- a screw-top jam jar with ridges round the neck
- a long piece of string
- bait (stale bread crumbs soaked in water work well)

1 Tie the string tightly around the neck of the jar so that it doesn't slip off. Tie a support brace around the bottom of the jar for extra security.

2 Fill the jar with water and place some pre-soaked breadcrumbs inside. For a deluxe trap, glue a little net bag to the bottom of the jar and put the bait in that. Lower the jar into shallow water where you can see lots of small fish activity and wait.

Step **2**

3 The idea is that curious fish will start to assemble around the jar, bumping it from all angles and soon finding their way around to its mouth. Once they have entered the jar, simply hoist it – water, fish and all – as quickly as you can to the surface. Once you have observed your catch, return it to where it came from.

This is all very well, you may be thinking, but why go to all this trouble when you could just push a net into the water and scoop them up? Well, there are several answers to this. First, small, open-water fish are not all easy to catch with a net – most have quite a turn of speed and the acceleration of a Ferrari! And they take some serious outwitting, especially when there is no cover. Second, jilly-jarring trains the amateur naturalist in the arts of observation and patience. Sitting quietly by a stream side waiting and watching the fishes' behaviour allows you to understand them better. And third, it is less destructive to the underwater habitat than even the most carefully wielded net.

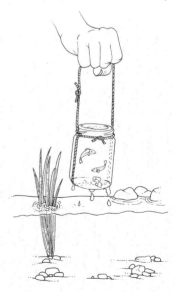

Step **3**

Fish-watching

Unless you get into the realms of huge amounts of money and expertise, aquariums have their limitations. For a much more complete experience, full of surprises, adrenaline buzzes and genuine discovery, try fish-watching. This can be as varied a pastime as the fish themselves: sitting on a cliff trying to spy the 1.5m (5ft) dorsal fin of a Basking Shark, wandering rockpools to find fish stranded by the tides, watching Minnows dashing in the shallows of a brook, chumming for sharks or donning the latest scuba gear and entering their realm.

Reading the ripples – passive fish-watching

Because they inhabit what to us is such a strange world, fish can seem very distant. Although we pass them every time we walk along a river, canal, pond or beach, we are rarely aware of them. But, like all wildlife, fish do occasionally give clues to their whereabouts, whether it is a splash, ripple or rise. So begin your fish-watching by using your ears.

An old Irish proverb, 'Listen to the sound of the river and you will get a trout', says it all to me, conjuring up a placid summer's evening listening to the soft kiss of Trout rising to take emerging mayflies from the water's surface. At the other end of the spectrum, fish can also produce a surprising amount of noise. Take my experience in Guyana with a fish known as a grunter. To this day I wouldn't know a grunter from a guppy, but I will never forget the sound – like a pig chewing pellets – that came up through the bottom of my aluminium canoe that day – an introduction to the acoustic world of fish.

Making water babies: fish lose inhibitions

Having said that fish tend to remain hidden in their secret world, at certain times of the year some species are so taken by the urge to breed that they lose their usual caution and drift closer to the interface between water and air. Procreation is top of their agenda. On a still spring day, look out for the violently thrashing water that signals the spawning orgies of Common Carp (*Cyprinus carpio*) – with other things on their mind, these normally timid fish become tolerant of human viewers approaching close, and what you see can be spectacular. The water around reedbeds fizzes with small cyprinids such as Roach (*Rutilus rutilus*) and Rudd (*Scardinius erythrophthalmus*) as males chase males and pair with females.

Brown Trout embryo development.

In the mating season, the schoolboy's 'tiddler', the Minnow, and its relatives go through a metamorphosis, with the males developing white tubercles on the head and gills and their bellies flushing pink, while the breeding females swell with their bodies full of eggs soon to be strewn on pebbles and stones on the stream bed. These frisky little fish seek the spring-warmed shallows; look for them congregating in large shoals for spawning. You can get good views of them if the water is clear and calm, but this is where a neat trick known as a 'jilly jar' (see p.146) can show you what pretty fish they are, with a variety of different colours from bronze to white with dark banding.

Lift stones from the stream bed, and it is not uncommon to come across a mass of pink eggs attached to the stones themselves or to the river bed. Nearby, unless it has been disturbed, will be an attentive male Bullhead, or Miller's Thumb (*Cottus gobio*). This is a small, secretive but common European river fish which seems to be known only to members of the angling fraternity, who occasionally snag one by mistake, or to kids who dabble with nets in fast and stony sections of stream or river.

Whatever you think of the Lumpsucker in the looks department, there is no denying that, if you bump into the brick-red-bellied male while he guards his brood of eggs, you will never forget the moment.

Another doting father who is not prepared to leave his pride and joy to fate is one of our most stunning sea fish: the Lumpsucker (*Cyclopterus lumpus*). Some may also argue that it is also one of the ugliest. This fish is large – some 40cm (16in) or so from head to tail tip, and rather rotund – which adds to the experience. The male is most likely to be found in shallow water among the kelp zone, where he has a brief rendezvous with the dull-coloured female before she leaves him to guard and ventilate the clutch of 30,000-odd eggs. While performing his loyal paternal duties, he may be caught out by very low spring tides and stranded as he will not abandon his brood.

One of the most bizarre fish, which has close relatives all around the world, is the Cornish Sucker, or Clingfish (*Lepadogaster lepadogaster*). This little rock-hugging creature, only 7cm (3in) long and extravagantly coloured in reds and greens, with two large blue eye spots and a mouth prolonged into a snout, can sometimes be found at low tide guarding its batch of yellow, skittle-like eggs, which are stuck to the underside of rocks. The best way to see it is to use a mirror mounted on a stick; this allows you to view the undersides of rocks and ledges, the sort of places these animals like to hide out.

The Cornish Sucker's real claim to fame is the device that enables it to live in the wave-battered world of the rocky shore – a suction disc on its belly formed from fused pelvic fins. This really works. If you catch one and try to pick it out of your net, it will sometimes stick to your palm and, no matter how vigorously you try and shake it off, it won't budge.

Of all the small fish spawnings, probably the most mysterious and bizarre are the writhing spaghetti masses of Brook Lampreys (*Lampetra planeri*) found in shallow stony rivers and streams in the spring. A lamprey looks a bit like an alien mutant eel, with a long sinuous body, a sucker disc of a mouth and seven gill holes that make you think it's been spiked in the head with a fork. The spawning is the culmination of a five-year life cycle. A gang of juveniles, known as a pride, hoover the stream bed for detritus; then, as they turn into adults in the autumn, their guts dissolve and their last action before they fade out is this mass spawning.

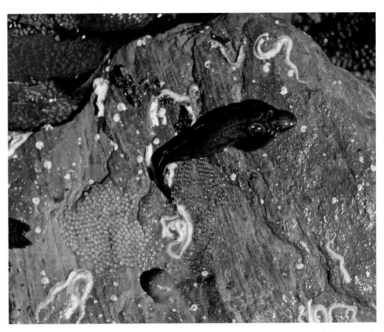

The Cornish Sucker or Clingfish.

Dummying fish

I cannot make my mind up on this one. Is it fish-watching or fish catching? Actually I guess it is somewhere between the two – it is angling without the intention of catching anything! Dummying is a way of instigating spectacular feeding behaviour from a fish, just enough to learn something before they wise up and let go. It can be tried with many different species. I have used fishing flies with the hooks taken off to encourage a Trout to strike at the surface and rise. Dragging a decoy rubber fish or even a dead one past a pike's lair will sometimes attract the interest of the resident and you will, by performing a bizarre puppet show, be able to witness the slow, 'I'm not really a predator' stalking behaviour as the animal almost imperceptibly leans towards its prey before snapping.

Feeding the fish – using baits as lures

A few bolder – or should that read greedier? – fish species can be attracted to baits and provide a refreshing addition to the fish-watcher's repertoire. Starting with the smaller fry and working up to the big stuff, you can liven up many a rockpool ramble by tying a piece of bacon or ham to the end of a fishing line and jiggling it to entice residents from their homely crevices. I have used this particularly successfully with the universal family of blennies, or Shannies (*Blennius pholis*), as these aggressive little intertidal scavengers are always up for an easy meal. You may even be able to pull them out of the water as they latch on to the bait with their many sharp teeth. They can be very territorial, so keep an eye open for interaction between competing individuals.

Shanny, or Blenny, as it's also known.

I have used similar methods with other fish species too. Red-bellied Piranhas (*Serrasalmus nattereri*) in the tropics of South America can be lured to some meat on a piece of string, and I have been able to witness the efficiency of their infamous teeth and jaws. A bunch of lettuce leaves on a string can be used to tempt vegetarian marine damselfish out into the open. Tying a bundle of worms together on the end of a piece of string, a practice known as patting, is a great way to attract and even catch eels, as is tying a small mammal carcass such as a Rabbit to some rocks in the clear shallows, sinking it into the water and returning after dark with a torch.

Probably one of the most spectacular takes on feeding fish, and about as far removed as throwing Koi pellets into my grandparents' goldfish pond as you can get, is 'chumming' for what has to be the most impressive fish in the world, the Great White Shark. The feeling of anticipation is almost unbearable as a bottle of clear, pure fish oil is dangled over the side of a stationary boat and a hardly perceptible trickle exits via a pin hole to set up an oily slick. This travels on the surface of the water, being blown away over the ocean by the winds.

To the sharks, not averse to a bit of scavenging, the slick suggests that there is food in the water. They pick up on almost impossibly small concentrations of the oil and, like dogs following the scent of a Rabbit upwind, work

The principle of using bait to attract fish works on big and small species alike.. The controversial practices of 'chumming' to lure sharks is on the bigger and more spectacular end of the spectrum.

their way along the trail, hoping to get an easy meal. Of course, feeding on a regular basis a fish that already has enough of an image problem and doesn't need to learn to associate boats with food is not a good thing, so those on board got to see our largest predatory fish swim a couple of cautious laps around the boat before vanishing as quickly as it arrived, leaving the dumbstruck fish-watchers with a sense of mystery and awe. The experience was not *Jaws,* but something far more lasting.

Feed with care

Using this sort of practice as an occasional embellishment to your fish-watching will probably have little long-term impact on the creatures or their environment. You should be aware, however, that feeding any wildlife continually may have complicated repercussions. The regular feeding of fish on coral reefs and at certain shark-watching and diving sites has already thrown up many questions and is frowned upon by some scientists. Use your discretion and bait wisely.

Reading the rings – ageing fish by their scales

The 13-century Holy Roman Emperor Frederick II had the right idea. He had a Pike which he 'tagged' by threading a ring through its gill covers. The fish was then released – and caught again in 1497, some 267 years later! The legend goes that it was a monster of some 1,350kg (3000lb) and almost 6m (19ft) long. All this sounds rather fantastical and a bit of a 'big fish story', but recent advances in science suggest that there may be an element of truth in it.

Identifying individual fish is possible. The big scales on a Mirror Carp, the bars on the flank of a Pike, the blotches on a Koi carp and the spots on a Trout are all as unique to that fish as a fingerprint is to us. This fact, combined with observations on captive animals and pets, has given us a basic awareness of what ages fish can attain. It is known that certain species can live for a long time; there is an unconfirmed (and questionable) report of a Koi Carp (*Cyprinus carpio*) reaching a stately 228 years; Spiny Dogfish (*Squalus acanthias*) can certainly live for 40 years, maybe up to a hundred; Lake Sturgeon (*Acipenser fulvescens*) have been recorded at 82; and there is an official record of a female European Eel (*Anguilla anguilla*) that reached 88 years of age!

Look closely at the scales of fish (particularly the large operculum that covers the gills) and you'll see growth rings. This one is from a Sea Bass and you can estimate its age by counting the growth rings.

But what if you want to tell the age of a fish that you haven't known all its or your life? Well, there are various ways. Depending on the species, otoliths (ear stones), fin rays and opercula (the bony flaps covering the gills) can be used, but for the most part your fish needs to be dead to deliver the required information. With living fish it is generally the scales that provide the answer.

Most fish keep growing throughout their lives. They also have scales that coat and protect their bodies, a little like a suit of chain-mail armour. Now the useful thing is that the pattern and number of these scales remain the same throughout the fish's life, so as the fish grows the scales have to grow too in order to maintain total body coverage.

The scale-based technique for ageing a fish assumes that growth is not constant throughout the year; it works particularly well if there is a seasonal aspect to the water temperature and hence to fish activity. Scales tend to grow fastest during the spring and summer, as the fish are feeding and the water is relatively warm. In winter the opposite is true and growth slows down. By looking at a fish scale under a microscope you can count the number of growth checks in very much the same way as you can count the rings on a saw-cut tree trunk: one check represented by a line or ring represents one year of age. What is also handy for ichthyologists (fish scientists) is that you can remove a scale and it will grow back with little or no negative long-term effect on the fish.

To scope a shark

A telescope isn't just a great investment for birdwatchers; it is also handy for watching fish – admittedly big ones! The Basking Shark (*Cetorhinus maximus*) is often best seen using a telescope (see equipment p.27). The technique is simple: find a piece of land that juts out into the sea in the west or north of the country and sit on it. You increase your chances considerably if you choose a calm day, as this creates the 'millpond' effect on the ocean that sea-watchers love.

Just sitting and watching makes for a thoroughly pleasant day, as even if the fish don't show there are usually loads of plants attracting large numbers of insects, and you are also in a good position to witness seabird movements or to catch a glimpse of the blue bullet form of a Peregrine. Headlands and bays seem to create excellent conditions for

Seeing the second largest fish in the world like this is difficult, but a view of one can be had by doing some sea-watching in the right place at the right time.

A Basking Shark, as you would be likely to see it when shark-watching.

plankton, either carried to the surface by water welling up from the deep or enjoying the shelter and warmer water these areas provide. Plankton is fodder for anything from a 10m (33ft) Basking Shark to the smaller fish, which in turn attract marine mammals that are worth looking out for at the same time.

Pondering over the sea requires concentration, as it is easy to become hypnotised or daunted by the sheer volume of water stretched out in front of you. But a careful combination of scanning with the naked eye and with binoculars or a telescope should enable you to pick up any discrepancy in the smooth ocean surface caused by a shark's fin. It is important to keep a wide view at this point, as is not unknown to be gazing intently out towards the horizon while a

Extra fascination: seabirds, fascinating to look out for and watch in their own right, can also give a clue to the whereabouts of fish shoals and other species that feed on them.

family of Bottle-nosed Dolphins plays unobserved at the bottom of the cliff below you. You are looking for clues such as a shining flank catching the light or what looks like a wave breaking or even heading in the opposite direction to the current. Watch the bows of boats for dolphins riding the pressure waves and large 'snowdrift' flocks of seabirds such as Kittiwakes or Gannets, which often indicate fish near the surface and a good focal point for dolphins or Harbour Porpoises.

Superfish

It is rare that a wild fish attracts a crowd of interested bystanders, but there they all were, locals out for a Sunday constitutional, dog walkers and dogs, screaming children in pushchairs; suggesting that whatever it was they were watching out for was either deaf or so engrossed in what it was doing that it simply wasn't going to be disturbed by all this racket. And this turned out to be true, because what the people were watching were fish – Atlantic Salmon (*Salmo salar*), called by some the 'king of fish' – taking part in one of the most ambitious odysseys in the animal kingdom.

By the time they reach the estuary that was the scene of all this excitement, the fish have already come a long way, travelling downriver from their spawning grounds and out to sea, where they spent several years fattening up on the ocean's bounty. Then on some cue known only to themselves, they head back to the mouth of that same river – a journey that has already involved finning several thousand kilometres! This is the last stretch, the home strait as it were, when the fish leave the relative obscurity of the sea and embark on a hormonally driven dash to the gravelly, well-oxygenated headwaters that are their spawning grounds. In fresh water they become ecological superheroes, supporting hugely important fisheries and sport for humans, and nourishing the many animals that prey on them, from bears to Bald Eagles. They have even been described as river fertiliser, as their bodies break down in the aftermath of their orgy and animals that drag their carcasses from the stream also provide a forestry service, enriching the soils in the river catchment areas.

The best thing about the Salmon spectacle is that it is very observer-friendly and relatively easy to predict; first you need to find a Salmon river with a good, healthy population. Some rivers support a spring or autumn Salmon run; others manage both. The conditions the fish are waiting for occur shortly after heavy rain or when the snow melts in spring and the river goes into spate. The Salmon will have been queuing in the lower reaches for the seasonal freshets or spates to lift water levels high enough to carry them over obstacles, smoothing out the harsh rapids, weirs and waterfalls where the best fish-watching is to be had. Here, the power of these sometimes 1m (3ft) long fish provides some remarkable athletic feats as they fight against the rushing currents. Rises in river levels may, however, be short-lived, and so progress upstream may consist of lots of stops and starts.

Where to look? Apart from weirs and waterfalls, any bottleneck in the river's flow is good. On quieter days the Salmon can be seen sitting

Little can beat the athletic determination of a salmon (here an Atlantic Salmon in Wales) heading upstream to spawn.

in pools below such obstacles waiting for the water to rise; they often stack up in slow, fin-flicking holding patterns, conserving energy until it is really needed and then unleashing it against the flow of the river. The best way to view them in these pools is to steal a technique from the angling fraternity and wear polaroid sunglasses, which cut out the reflections and allow you to see 'through' the surface of the water.

Although watching Salmon leap up 3m (10ft) waterfalls is spectacular, the best bit for me is following them right up to their final destination, the culmination of all this effort – the gravel beds where the spawning occurs. Here the water is clear and alive even when the land is locked in the frosts of winter; you can watch the cocks jockeying for females and position with much splashing and foaming of water, and then see them 'chaperone' their chosen mates, jealously guarding them from the attentions of any other males. The females then start thrashing the gravel with their bodies, scooping out the nest, or redd or hollow in which they lay their thousands of eggs. All this action takes place in water that can be so shallow the fishes' backs break the surface, and the charges of the males who have not yet attached themselves to a female can be so vigorous that some individuals get carried away and run aground. To watch the very end and beginning of this iconic fish's life cycle is well worth any effort involved in getting to the site.

Atlantic Salmon eggs and fry with yolk sacs.

Battledress

Depending on where you are watching your Salmon and how fast they are making their journey, they will seem very different in appearance; those that have just left the ocean will be silvery in colour, but as they head upstream many physical and physiological changes take place. These are most obvious in the cock fish – not only do they stop feeding, but they also take on battledress; their flanks blush with all the colours of the rainbow and their jaws distort into

Male Atlantic Salmon showing breeding colours.

a hideously warped warclub to be used against other males: the jawbones lengthen to form a hook, known as a kype, and the teeth grow longer. At the same time their gut shrinks and their reproductive organs develop.

~ HOT FLUSH – SPRING TIME ~

The robust and feisty Three-spined Stickleback (*Gasterosteus aculeatus*) can be found in saltwater, brackish and freshwater habitats pretty much throughout the northern hemisphere. It grows no more than 10cm (4in) long, but what it lacks in stature it makes up for in intrigue.

Male Sticklebacks become very territorial in the spring when they feel the surge of their hormones; they blush with a crimson belly, and the normally silvery eye turns bright sky blue. Set up a 60cm (2ft) aquarium and let it settle down for a week (see p.142). Provide plenty of detritus in the form of pond weed, dead leaves and a little sediment, preferably from the pond from which you will be sourcing your Sticklebacks.

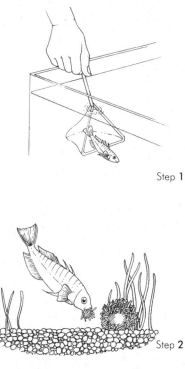

Step **1**

You will need:

- a small fishing net
- male and female Sticklebacks
- an aquarium already set up
- mirror
- cardboard
- scissors

Step **2**

1 Find a male Stickleback by fishing in a pond with a net on a warm spring day, and put him into your aquarium. Leave him for a week to settle in, feeding him on small pond creatures or bloodworms, available from most aquarium and pet stores.

2 Before too long you should see the fish sucking up mouthfuls of mud and arranging weed and detritus on the bottom of the tank. He is building a nest and will fuss over it with great attention to detail.

Step **3**

3 Place a mirror against the glass, and your fish should go into an aggressive display, thinking the reflection is a rival male in his part of the pond. He will attack it with his mouth open and spines erect.

4 Cut out three Stickleback outlines from cardboard. Leave one white, colour one red and give the third just a red belly. Place these against the glass. Which one does your Stickleback attack most vigorously?

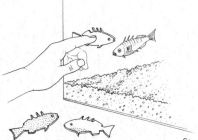

Step **4**

5 Then go out and catch a fat female Stickleback who is full of eggs. The females look like the males but are an olive-green colour and lack the blue eyes. Introduce her to the tank and, assuming you have got it right, you will see a zigzag courting dance. If you are lucky, the female will be suitably impressed and the male will lead her into the nest. Here she will lay her eggs and he will fertilise them with his milt.

6 When all this is over, remove the female and in a few weeks the doting male will fan, aerate and guard the eggs, before becoming a proud father before your very eyes. Remember, when you have learned as much as you want from your fish, you should release them in the water from which they came.

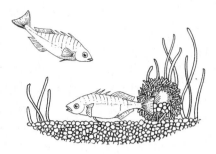

Step **5**

Step **6**

Red and you're dead

The world-famous Dutch ethologist and 1973 Nobel Prize winner Nikolaas Tinbergen kept a fish tank like the one described above in his laboratory. One spring he noticed that, at a certain time every day, his captive male Stickleback would start swimming in an excited manner and attacking the glass of his home. It didn't take the astute Tinbergen long to make the association between his agitated fish and the movements of the red postal van that drove past his window every day at the same time. The fish was responding to the colour of the van, which was the same as the belly of a rival male. This association between what are called sign stimuli and the resulting behavioural reflexes of the fish led to a series of very neat and beautifully simple experiments, on which the above project is based.

In certain lakes in North America, some male Three-spined Sticklebacks have a black belly and throat instead of the normal bright red one. It has been found that, although they have a little more difficulty impressing the ladies, they live longer, as their black throat doesn't attract predators in the same way as the red.

The pugnacious and often overlooked Three-spined Stickleback.

Looking through the mirror

One of the problems of being a terrestrial animal trying to peer into the underwater world is that the interface between the two – the surface of the water – reflects light. In fact it does this in more than one direction; anything but the smoothest, stillest water has a surface that is being interfered with, rippled, tugged, torn and pulled in all directions by the wind and currents below. This means that light bounces in all directions, so that no matter how clear the water is, looking through it can be difficult, if not plain impossible.

Fishermen get round this problem and reduce the glare on bright days by donning polarising sunglasses. These have a special coating that reduces the amount of reflected light. As light comes down from the sun, hits the water surface and bounces in all directions, the light that is flung off in the horizontal plane is what is known as 'glare'. Polarising glasses cancel out this glare, leaving just the ambient light and allowing the wearer to see through the surface. It's as simple as that. When I was little, the only good polarising sunglasses were huge and made you look like Elvis, but nowadays they come in a variety of styles ideal for the fashion-conscious fish-watcher!

Looking cool, fisherman know this trick or two, a pair of polarising sunglasses cuts out the glare allowing you to see what's beneath the surface of the water.

~ UNDERWATER VIEWER ~

Another way to see beyond the confusion of the surface is to use this simple device, which is equally effective in salt and fresh water. It works on the principle of pressing a clear window against the water, cutting out ripples, while a baffle cuts down the glare and improves viewing. The beauty of the design is that, if the window is damaged, you can replace it very easily.

Step **1**

You will need:

- an ice-cream tub with a close-fitting lid
- scissors or a knife that will cut through the tub
- plastic waterproof paint – black
- clingfilm or clear cellophane

1 Cut the bottom out of the tub and the centre out of the lid, leaving the edge with the seal to act as a frame.

2 Paint the insides of the tub to eliminate glare.

3 Stretch the clear film over the top of the tub.

4 Put on the lid to fasten the film in place, press the clingfilm against the surface of the water and start viewing.

Step **2**

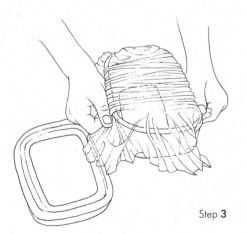

Step **3**

Step **4**

Total immersion technique

snorkelling

My first experience of this ultimate underwater-viewing experience was lying on the rocks with my face encased in a mask immersed in the shallow water of an intertidal pool. It was just a matter of time before I cast off my land ties and started experiencing the freedom of getting as close to being a fish as a human can.

All you need to get started are the regular mask, snorkel and fins, but a wetsuit is a good idea in temperate zones, as the cold can really put the kibosh on your enjoyment and limit your time in the water. Once you are kitted out, a whole new vista opens up. Animals that cannot tolerate the intertidal zone are just a fin stroke away, and the diversity of life to be seen is extraordinary. I remember being totally blown away by the colours and the curious creatures that I hadn't even heard of, and that was just below the surface off a pier in a south-of-England holiday resort.

We snorkel in the sea so why not in a river or pond? So long as the right precautions are made this can be a surprisingly rewarding occupation.

A fresh approach to fresh water

It's a strange thing, but although most of us wouldn't think twice about dressing up in fins, mask and snorkel at the seaside, donning the same gear in the course of a riverside walk seems tantamount to social suicide. This is a shame, because river snorkelling can be every bit as rewarding as ocean snorkelling. The only real differences are that your buoyancy is less in fresh water; and in some rivers, where there may be obstacles and little space for manoeuvring, fins become more of a hindrance than a help.

Maybe the reason river snorkelling hasn't caught on is that it is just too extrovert. Don't be surprised if you are greeted by a crowd of less enlightened observers. Most are just curious about what you are doing; those with richer imaginations may think you are a police diver looking for a body. I remember feeling a little self-conscious the first time I put on my suit of black crushed neoprene and slipped into my local river. But after a few minutes' observing Trout lying up in the eddies behind rocks, hearing the constant fizz and bubble of the moving water in my ears, watching frisky Minnows chasing each other in the shallows and even being allowed in the deeper water to hold court with the king of fish, the Salmon, any embarrassment about what I was doing or looked like dissolved into the watery world in which I was immersed. The experience is so addictive that I often leave the water after an hour or so, having lost most of the feeling in my extremities just one short shiver away from hypothermia. But never mind that – this is the only way really to enter the underwater world. There are no words to describe what it's like – I just want to encourage everyone to share it. Come on in, the water's lovely.

Snorkelling with a beast. Here spending time with a Wels Catfish in its natural habitat is a much better way of understanding the way the fish lives.

A few safety tips

I do not want to scaremonger here, but entering any underwater environment is potentially hazardous. As is crossing the road, of course, but we take those rules for granted as we have known them all our lives. And with the water, as with the road, the precautions are just common sense once you think about them.

It was a scene like this – salmon and trout in clear water that looked so inviting – that lured me into the water in the first place – I've never looked back since that day.

Go on a course, even if you are not intending to use scuba equipment. Snorkelling and scuba diving share many skills and techniques, and scuba diving in itself is a very useful skill for the naturalist, but if this feels like overkill, there are professionally run courses that focus on snorkelling. Make sure you can snorkel in a controlled environment before taking the plunge into wild water.

Choose your water wisely and avoid swimming in fast-flowing water, locations that are subject to strong currents or flash floods, or anywhere upstream from a weir, rapid or waterfall. It is all too easy to be swept away.

Buddy up. An experience shared is an experience doubled, as they say. It is also sensible to swim with a partner in case one of you gets into difficulties. It is twice as easy for two to get out of trouble as one.

Keep a look-out for predators in the ocean and in some freshwater habitats, especially in the tropics and subtropics. I realise that getting close to wild animals is what it's all about, but there are some where a little distance is preferable. I have been surprised a couple of times by a crocodile as I have tinkered around in rivers in Africa and South America, forgetting that I'm not at home in England, where such precautions are unnecessary. So do your homework first.

Spineless wonders
Invertebrates

An alien returning to its spaceship after a visit to earth would report back to its leader that earth isn't dominated by a race of clever apes bent on self-destruction, but by a huge number of strange, spineless animals. They live almost everywhere, from the skull-crushing depths of the oceans to the peaks of the highest mountains, from burning deserts to the frigid poles. And the alien's report would identify them as the most important organisms on earth.

Those without a vertebral column, or spine, rule the planet, not once upon a time, not after a nuclear holocaust, but now, today. Take the time to look and you'll find them everywhere your eyeballs rest. Scale down your perspective and you'll find them in the lush forests and grassroots jungles of a field or lawn, roaming the deserts of the patio, and as aeronauts and aquanauts taking death-defying chances in the sky and underwater. Even if you don't notice them, you'll certainly be aware of their effects on our world, both positive and negative.

The first arthropods were scuttling around on the planet's surface around 500–600 million years ago, and their soft worm- and jellyfish-like ancestors were oozing around the world's oceans and wet places long before that. Over this vast expanse of time these animals have perfected their craft. They are integral cogs in the machinery of every ecosystem on the planet. If you wish to see beauty, adaptive forms and specialisations that you couldn't even begin to dream up, look to the denizens of planet bug and you'll find your inspiration.

About 97 per cent of animals are invertebrates. If you were to sift and sieve through a 1-metre cube of soil you could well find between 500 and 2,000 insects alone. That's 10,000 million per square km of land, or 200 million insects for each person on earth! Without them there wouldn't be flowers and therefore fruit, nor any of the fluffy and feathered animals we love so much, and without their composting and recycling services we would be knee-deep in dead leaves, dung and dead animals.

No bones here – a snail is a model invertebrate. With no spine, just a simple soft body and a shell to hide it all in, giving it protection from predators and the environment.

Left: Representing probably the single most successful group of invertebrates, the beetles. They don't come much more spectacular than members of the stag beetle family. This is a male European Stag Beetle demonstrating the extreme designs that the beetles' exoskeleton can be turned to.

Bodies in a box

What is an invertebrate? This word translates as 'without a backbone'. Most invertebrates belong to a further subdivision or phylum called Arthropoda – the arthropods. All arthropods have an exoskeleton made of a remarkable substance called chitin. Their bodies are effectively enclosed in a box, supported from the outside (in contrast we have our supporting structure – our skeleton – on the inside; it is called an endoskeleton). Their limbs and bodies are jointed and segmented; the word 'arthropod' means 'jointed leg'.

Recent estimates have suggested that there are between 25 and 30 million kinds of arthropods in the world, most of them yet to be discovered. The major arthropod groups are the insects, chelicerates and crustaceans.

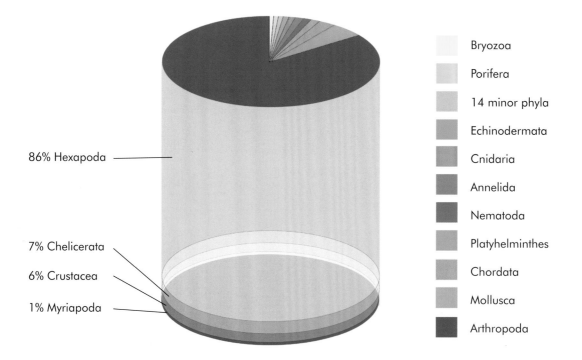

- Bryozoa
- Porifera
- 14 minor phyla
- Echinodermata
- Cnidaria
- Annelida
- Nematoda
- Platyhelminthes
- Chordata
- Mollusca
- Arthropoda

86% Hexapoda

7% Chelicerata

6% Crustacea

1% Myriapoda

This 3D chart is intended to show how vast the arthropod group is compared to other animal groups. Within the arthropod group there are 4 main subphylum shown here in the chart down the left side, see opposite for further information about each subphylum.

Six of the best – hexapods
(Subphylum: Hexapoda)

Almost all hexapods are insects (class Insecta). This class is divided into about 30 sub-categories or orders, some of which we will explore in greater detail later. There are about one million hexapods known to science. They all have six legs (the Greek word hexapoda means 'six legs'), their bodies are divided into three parts (head, thorax and abdomen), and their breathing system is a network of hollow tubes called trachea.

House Fly (*Musca domestica*)

Horny claws – chelicerates
(Subphylum: Chelicerata)

The other major arthropod group comprises the wonderfully freaky-looking chelicerates, or 'claw-horns'. There are about 80,000 species of arachnids (class Arachnida), a group of eight-legged animals that includes spiders, scorpions, ticks and mites. Other chelicerates include the horseshoe crabs (class Merostomata), which are not true crabs, but more akin to spiders, and the mysterious sea spiders (class Pycnogonida).

Dog Tick (*Ixodes scapularis*)

Crabs and kin – crustaceans
(Subphylum: Crustacea)

Another big group of arthropods, Crustacea includes the well-known crabs, lobsters, shrimps and prawns, but also barnacles, triops, water fleas, krill and remipedes. Most are aquatic although among the 30,000 or so species are a few that have managed to invade the land, such as the woodlice and various land crabs.

Land Crab (*Johngarthia lagostoma*)

Millipedes, centipedes and allies
(Subphylum: Myriapoda)

There are around 15,000 species of these, split into four further classes, the most frequently encountered of which are the millipedes (class Diplopoda) and centipedes (class Chilopoda). One thing they all have in common is an elongated body with many legs running along its length.

African Giant Black Millipede (*Archispirostreptus gigas*)

Other invertebrates

The remaining invertebrates are a motley collection of mainly soft-bodied animals, and among them are the largest invertebrates in the world, the giant squids. They include molluscs, true worms, flatworms, and a range of other groups.

Dog Whelk with eggs on a rocky shore.

The 'Bug Effect'

It is a pity that many humans fear the creepy and crawly forms of life, especially since we ALL need them. It never fails to amaze me how many people respond badly to them, with reactions ranging from backing away making polite excuses, to uttering guttural noises of disgust, or even running away screaming with fear! Be patient and keep in mind that very few people are truly phobic about these things. A true phobia is an intense response, as if in a genuinely life-threatening situation. Symptoms include a quickened pulse, shortness of breath, cold sweat and faintness, followed by a flight or freeze response. All of these manifestations would not be out of place if you were charged by a rhino, but with small harmless spiders it really doesn't seem logical.

The naturalist working with invertebrates and other 'phobia-generating' animals such as snakes and frogs should expect these negative reactions, but try to slowly work in some positives. Never force anyone to come to terms with their fears — most people will be a lot better off if they arrive at this point at their own speed. And of course be prepared to admit failures — some people are simply too far gone to convert to bug love!

Banded-legged Golden Orb-web Spider in web, Gorongosa National Park, Mozambique.

Bug-hunting essentials

The best thing about invertebrates is that they are everywhere. Wherever you are in the world, they are never far away. They easily capture the attention of the curious and open-minded; the person who is willing to get down on their knees and take a closer look will never, ever be bored.

When I was a kid, a Great Diving Beetle (*Dytiscus marginalis*) that landed on the bonnet of my parents' car in a supermarket car park saved me from impending boredom. But that moment also showed me how pond insects colonise new habitats, spotting ponds by the polarised light reflected from them. Shiny cars unfortunately do the same thing, and to a water beetle a car park looks like a lake.

Insects have often entertained me in dull situations. Stranded at an African airport for hours, a mud-daubing wasp was about the only air traffic moving. Soldier crabs and their sand-bobbling actions were the perfect cure for seaside resort boredom, and a foraging line of ants on a hotel toilet wall took my mind off the stomach cramps.

A Great Diving Beetle, a pond dipper's dream, can turn up anywhere.

Another great thing about invertebrates is that you require very little specialist equipment to study them. Most of what you need can be made from recycling other household objects. While the word 'microscope' implies 'expensive' (and many are), even this luxurious portal into the microcosm of invertebrate life is becoming more affordable. A very good microscope, which can be plugged into a USB port on a computer, can be bought for the same price as a video game. An excellent beginner's microscope can be acquired for twice the price of a video game console, and I guarantee will deliver just as much excitement, while at the same time being 100 per cent educational!

That said, you can still travel a long way into the invertebrate world armed with nothing but a hand lens. With this, you can meet the gaze of 36,000 eyes, observe virgin birth in your herbaceous border, and witness the trials of the grassroots jungle first-hand. I've spent most of my life specialising in the minutiae of life, and besides the pocket hand lens I would also not be without a notebook and/or camera with macro facility, and plenty of specimen tubes. Additionally, a net and a pooter are handy at times.

A pocket loupe is an essential bit of kit for bug-hunting; here it's being used as a scale against a Scorpion.

A word on the ethics of entomology

Just because they are small, and are lunch for most of the rest of the natural world, that's no excuse to cast your usual ethics aside as a naturalist. Invertebrates get a rough deal from us humans as it is. Always remember that they are living things that need to be respected. If you catch some for observation in captivity, always release them where you found them afterwards.

Whether looking at them for a few minutes, hanging onto them for an hour or two, or keeping them for a longer period to study, they are living things like you and I and therefore deserve as much respect. Here a spider resides in an arachnid 'Hilton'.

Look at it this way

The true joy of being a naturalist does not lie in owning expensive fancy gear, but in observing something incredible happening right in front of you. Our eyes are our most important tools as naturalists – the problem is most of us do not know how to use them, especially when studying very small animals.

You need to train your eyes to work for you. Lie belly-down in the grass on a warm summer's day and stare into the sward. You'll start off seeing very little, but tune your eyes in, and let them become accustomed to small movements. Soon you'll become aware of more and more, such as ants scurrying up plant stems, the shiver of an aphid being attacked by a parasitic wasp, the rasping action of a grasshopper's leg or the wing-flicking of small dancing flies on a leaf. What was just a patch of scruffy long grass becomes a veritable forest of life.

Now slowly stand up, look at the sea of grass stretching out in front of you, and then the scale of it all hits you. Suddenly the behaviour of other animals – the low flights of Swallows, the pounce of a Fox or the questing of a Badger's nose – all starts to make sense. By tuning your eyes in to the little lives, you understand more about the big ones too.

Parasitoid Wasp with Cypress Aphid. Look close enough and even the tiniest life is up to extraordinary things.

Pots of pots

I know we've mentioned them a lot already in this book, but pots and specimen tubes are important to the naturalist, and vital to those undertaking the study of invertebrates.

Many small animals are both fragile and need close scrutiny to appreciate or identify correctly, and this is where your collection of pots comes into its own. In the field, small pots serve as pocket-sized protection for delicate specimens, either living or dead, and are useful for observation and identification. Bigger ones, such as ice-cream or margarine tubs, modified with a few holes or some netting, make excellent habitats for slightly more long-term captive study.

Every cupboard around the house is full of containers that have some use to me. My favourites are Tic Tac boxes and old photographic film canisters (you can still get these from proper photographic shops but they are rarer now then they were). The clear plastic tubs that Chinese take-aways come in are also rather handy. Glass specimen tubes from a naturalists' or entomological supplier are inexpensive, come in a variety of sizes and are available in plastic or glass. Plastic ones won't shatter if dropped, but do tend to scratch and mark easily.

Another fantastic container for small creatures is a 'pill box'. These are small cardboard tubes, with a lid at one end. They have many advantages – they're lightweight, different sizes fit into each other like Russian dolls, they don't sweat, they breathe nicely, insects can grip the sides easily and deluxe models have a transparent plastic end, for observation. The only downside is they become useless when they get wet, but there are posh ones with a waterproof coating on the outside, or you can waterproof them yourself by painting the outside only with clear varnish or covering with duct tape.

You can never have too many pots and vessels – a good bag or pocket full should be carried wherever you go.

Some potting and tubing ethics

These pots and tubes are airtight, but there is no need to make holes in the lid, for the air within will last a small invertebrate creature many hours (if not days). The biggest danger to them is getting damaged by being physically tossed around or by drowning in their own 'sweat' – all living things give off moisture which condenses on the glass. Both of these problems can be addressed by popping in a strip of kitchen towel or toilet tissue. The bug then has something to cling to, which also calms it down, making it easier to observe. If you are keeping a vegetarian animal like a caterpillar or grasshopper for more than a few minutes, maybe also place a leaf of foodplant or blade of grass inside for them to nibble.

~ THE BUG RESTRAINED ~

When you study invertebrates, especially insects and arachnids, you'll soon find that one of the biggest frustrations is that most move very fast and erratically! Give the captive too much three-dimensional freedom and it will scurry around and up the side of the pot, making observation impossible. A 'bug restrainer' might sound like a torture device but trust me, it works really well and if used carefully doesn't harm anyone.

You will need:

- 2 plastic drinking cups
- a polystyrene tile
- marker pen
- scissors
- PVA glue
- clingfilm
- elastic bands or sticky tape

1. Cut the middle out of the bottom of one of the cups, leaving a rim. This is cup 'A'.
2. Draw around the base of the other cup onto the polystyrene tile. Then cut out this disc and glue it to the underside of the base of this cup (cup 'B').
3. Stretch a piece of clingfilm over the base and the hole in cup 'A' then secure it with tape or elastic bands.
4. Once the glue is dry your 'bug restrainer' is ready to use. Place your hyperactive bug in cup 'A' and then slowly and gently push cup 'B' into cup 'A' until the bug is gently sandwiched between the clingfilm and the polystyrene. You are now free to have a good look at your specimen. This speeds up the whole process between capturing, identifying and releasing – better than keeping the bug for a long time in a pot.

It may take a couple of attempts to get your specimen in the best position. Once it is pinned down, work quickly, and then release it as soon as possible. Always make sure your bug is dry; condensation will spoil your view and the bug will get in a right mess. And never try to restrain a squishy bug like a caterpillar, or one that looks like it will be bigger than the base of the cups when fully stretched out.

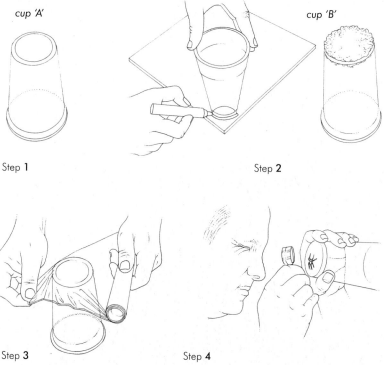

cup 'A'

cup 'B'

Step 1

Step 2

Step 3

Step 4

~ BUG-SUCKER – MAKING AND USING A POOTER ~

A pooter is a small, lung-powered vacuum cleaner, with which you can effectively and quickly suck up small fiddly invertebrates into a clear vessel. There are several different designs on the market, and while they are relatively cheap, a home-made one is just as functional.

You will need:
- small jar with a plastic lid (the kind you get herbs in are ideal)
- 50cm (20in) of clear plastic tubing, approx. 5mm (¼in) diameter (wine-making shops or DIY stores should stock this)
- coloured insulating tape
- elastic band
- small piece of muslin or any fine-meshed fabric
- scissors
- candle
- matches
- meat skewer
- modelling clay or similar

Step 1

Step 2

1. Take the lid off the jar. In a well-ventilated place, use a lit candle to heat the tip of the meat skewer. Use the skewer to melt two holes in the jar's lid, slightly less than the diameter of your plastic tubing. The tubes must fit snugly into these holes.
2. Cut the rubber tubing into two pieces, one 20cm (8in) long, the other 30cm (12in) long. Insert each into one of the holes in the plastic lid (run them under a hot tap if you are having difficulties, it makes the plastic softer); push them in so that they would reach about halfway into the jar.
3. On the shorter tube, put a piece of mesh over the end that will go in the jam jar – secure it with an elastic band. Wrap a small band of coloured insulating tape near the tube's other end – this is to remind you which tube to suck on. If you suck on the other one by mistake you could end up with a mouthful of bugs. Place the lid on the jar.

Step 3

4. Your pooter is now complete. I recommend practising on small pieces of paper or fluff before you try it on living things – give a short, sharp suck on the end of the short tube, while directing the end of the long tube over your target. When done correctly it will transport the desired small thing into the jar. The muslin on the end of the short tube stops you inhaling your catch!

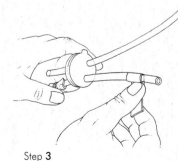

Step 3

Step 4

A guide to good 'pootering'

Only suck up small animals in your pooter. Anything fragile, with long legs or with a body nearly as wide as the tube will get stuck, get damaged or die. Try to keep the collection jar moisture-free. Don't blow into the tubes – this fills it up with moisture from your breath. Slimy animals such as slugs and snails are not good to 'poot' up as their mucous sticks everything to everything else. Keep the number of animals in the collecting jar to a minimum. Keep predators such as spiders away from other animals.

Always suck through the filtered tube. It's a good idea to mark it with some tape.

Fiddling with forceps

Forceps, also sometimes called tweezers, are useful things. They come in a range of different sizes for different jobs – super-fine ones for fiddly jobs; long ones for handling venomous creatures, feeding things with teeth or adjusting things in an aquarium; curved ones with broad tips for grasping entomological pins. I own and use all of these regularly!

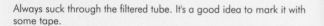

Marking a Honey Bee, by holding gently with forceps and applying a dab of quick-drying paint.

I have a few pairs I use more than the others: a regular set, for turning over nettle leaves, a very fine (expensive) pair for working under a microscope, and a lovely 'gentle' wide and flexible pair for very delicate operations, like plucking a spider skin from a web, and picking up living things. Most forceps are, without modification, too harsh for use on living things.

Through the loupe: hand lenses

There is magic in magnification. To appreciate the detail of a dragonfly's eyes or just how hairy a spider is you'll need a little optical help. A hand lens, pocket magnifier or loupe is probably the one indispensable bit of naturalist's kit. There are many types out there to suit different budgets but a good-quality metal one – ideally with a screw and not a rivet holding it all together – is best. (They do fall apart, so regular checking of the screw, or a little thread lock, will save the use of a piece of twisted wire further on down the line.) The beauty is that this little portal to another world can be easily pocketed or, with a bit of string, hung around the neck. A magnification of 5× or 10× is most useful. Some even come with multiple lenses.

To use your loupe, hold the lens between thumb and forefinger of your right hand (if right-handed), and bring the lens up to the left eye, resting the edge of your finger against your brow and the edge of your thumb against the bridge of your nose. This way the distance between the lens and your eye is fixed. Now bring your specimen up towards the left eye and magnifier until it is in focus.

A magnifying lens is a game-changing affordable luxury, allowing an amazing perspective on the Lilliputian world of invertebrates.

Lights please

You will be surprised at just how often you'll find yourself grovelling around in dark places. A small torch is handy for illuminating deep cracks and crevices. Failing that a small pocket mirror is also very useful for bouncing light into places where it's needed, but you do really need sunlight for this to work well.

You'll be surprised at just how useful a small torch can be, from looking under bark, into crevices in rockpools, even back lighting a leaf miner.

Net gains

We've mentioned the usefulness of nets already in the book, but it is when in pursuit of invertebrates, whether marine, freshwater or airborne, that a net really comes into its own. Nets are extensions of your body, allowing you to gently bag things which would be very difficult to catch with your hands and arms. But resist the temptation to start swinging your net around like a windmill. Only use a net if you really have to, for example to identify some restless winged thing that never seems to settle down, or to collect things out of reach, hidden in foliage or even underwater. There are three main kinds of net.

Butterfly net or kite net These are used to catch aerial creatures. They have a bag made out of a light mesh that is gentle on the wings of insects, although they do not fare so well when they catch vegetation, as the fabric tears easily. Those made with black material are best as they allow you to see your captives through the net – handy if you have unwittingly netted a bee or wasp!

Sweep net Strong nets supported by robust frames with a heavy and tough cotton bag, these are designed to be swept through vegetation and can take quite a hammering. Sweep-netting is a shotgun technique that nearly always turns up something interesting.

Dip net Another thick strong net, designed to be used in water. The main requirement is that the weave of the fabric isn't so tight that the water drains out too slowly, but is strong enough to resist abrasion by underwater obstacles and snags. There are many other forms of specialist nets for use in the aquatic environment, such as shrimp nets with a strong flat bottom profile, and nets with a canister sewn into the bottom of the bag – perfect for collecting tiny pond animals and plankton.

Horses for courses –
top right: course pond net
top left: sweep net
bottom left: butterfly net
bottom right: fine pond net

You can of course buy these from specialist shops but with a little effort, some sewing and a garden cane, you should be able to come up with something. A sieve or tea strainer on a pole makes as good a pond net as you could buy.

A word on nets

'Look before you swing' is a good code of practice for the owner of a net; it is all too easy to become a net-wielding maniac bent on bagging your bug. You'll learn a lot more about an insect flying free than one caught in folds of fabric. So resist the temptation and only put a net into play as part of a sampling exercise or when all manner of stealth and patience has failed you. The same ethos applies to the use of dip nets: peer between the lily pads and try and see through the water's surface before stirring it up into a muddy mess.

A few rules of engagement if a net is needed. With the correct use of a butterfly net and the right kind of observation pots, you can catch, bag and identify without even laying a finger on your subject.

Use your net gently and gracefully; picking the insects from vegetation or from behind in flight is best and as in tennis keep your eye on the 'ball' to avoid hitting the insect with the net rim, which at best stuns the animal and at worst will break it. Try and avoid swiping with the net as this can also damage the fragile wings. If the insect is flying, swing the net from behind if possible and as soon as the insect is in the net, with a quick flick of the wrist fold the net, trapping the insect in the bag.

Once you have your bug bagged you can view or identify it immediately through the net fabric. Lifting up the end of the net bag and allowing the insect to fly or crawl up is the best way of getting it into a useful position. Or you can transfer it to another vessel for observation. Cup the pot over the insect – pots with a clear bottom have an advantage here as insects always crawl toward the light. Then slide the lid on and you've got it.

A net is a useful tool but use it wisely. As soon as an insect is captured it stops behaving naturally.

Tickle and drop: spoon and paintbrush

Many insects and spiders have a really frustrating habit of dropping to the ground at the slightest disturbance. I'm sure some do it if you just look at them! Here's where a tablespoon comes in handy to catch the falling insect. Use a small soft paintbrush to sweep or flick delicate animals such as caterpillars off plants or the ground and catch them in the spoon.

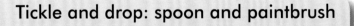

Preserving

Once upon a time, not so long ago, people who liked butterflies would catch and kill them, and stick them with pins. A strange way to show your affection for an insect, perhaps, but it was the age of the collector and that was what people did. Collecting butterflies and moths has a stigma attached to it now, and some people see those with a net as a butterfly collector.

Now, although I really do not condone collecting for the sake of a having a collection, I'm realistic. There is sometimes a genuine need as part of a study to form reference collections, especially in the understudied tropical regions where discovering new species is a very realistic possibility. It would be hypocritical of me to be completely anti-collecting, as it was a collection of moths that I made in an old cigar box that partly sparked my interest in insects as a small boy. Even if the humane use of a killing jar is not for you, insects can still be stumbled upon already dead and you may wish to preserve these instead.

Knowing how to preserve and display a specimen is a dying art in our modern world, but it has its place.

Freshly dead insects will be flexible and easy to manipulate, and will remain so for 24 hours. Most can be preserved by allowing them to dry out. Insects which are long dead will be dehydrated and rigid, but can be 'relaxed' again by reversing the process and exposing them to a humid environment. Traditionally this was done by putting your specimen in a 'relaxing tin' along with cotton wool impregnated with 'relaxing fluid' (containing the mould inhibitor Chlorocresol), available from biological suppliers. Water will work, but specimens can go mouldy. If you cannot get hold of the right stuff, improvise by adding disinfectants to your water.

There are standard ways of arranging different insects. Butterflies, moths, dragonflies and the like should be spread out to display all the wings at once. For this delicate and fiddly process you'll need pinning board, pins, a setting needle, entomological forceps and thin tracing-paper strips. All of these can be improvised, but a good naturalist will have invested in a proper pair of entomological forceps that are designed to manipulate and grip setting pins. Proper entomological pins are the best as they are made from stainless steel and do not corrode like dressmakers' pins. A setting needle is an inexpensive addition to your dissecting kit but can also be cheaply made by setting a regular needle in the end of a length of round wooden dowel.

With winged insects, you set the pin through the thorax, carefully and firmly, taking care not to damage the wings. The pin should be on the right side, close to the base of the wings. The insect should then be pinned into the groove of a setting board. This is made of a soft material such as cork or foam. Professional ones are covered in paper and come with different-sized grooves (5–20mm (¼–¾in) wide are standards) to accommodate different-sized bodies. I have in the past made my own using polystyrene or Balsa wood. Different-sized pins are used for different species of insects. Pins are of a standard length (3.8cm, 1½in) but come in a variety of thicknesses. Most useful are no. 5 for large insects, no. 3 for medium-sized insects and nos. 1 and 2 for very small specimens. Winged insects can then be arranged by dragging limbs and wings into position.

Specimens should be stored in a dry atmosphere and preferably in sealed boxes. It really depends on what you wish to do with your specimens as to how you store them; even cigar and chocolate boxes can be used as long as you place a block of naphthalene inside the storage case to protect them from damaging pest insects and mould. If you are keeping them out on display, try and avoid bright sunlight; many pigments are unstable and light makes them fade very quickly.

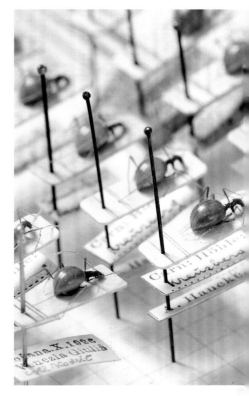

Slenderneck Beetle specimens, Postojna Museum of Natural History, Postojna, Slovenia.

The killing jar

I do not want to linger on this device but as I'm trying to be comprehensive here, it needs a mention. The killing jar is a quick, humane and efficient way of killing insects and certain other arthropods. It consists of an airtight jar with a wide mouth and a layer of plaster of Paris in the bottom. A few drops of 'killing fluid' (ethyl acetate), available from a professional supplier, is then dripped onto the permeable plaster and allowed to soak in. A few sheets of crumpled tissue paper should then be added to stop the insects getting damaged. An alternative is to use a similar airtight jar with a few fresh finely chopped Laurel leaves; these contain the lethal chemical cyanide and if left for half an hour or so before using the jar will build up in sufficient concentration to be effective. The insects should be placed in the jar and the lid quickly replaced.

Killing jar

NB: Many of the chemicals used to professionally kill, relax and preserve specimens are hazardous and shouldn't be used unless you've got experience or are doing it with an experienced person.

The insects

Insects rule nearly all terrestrial ecosystems. Wherever you are right now, look or poke around a bit and you'll see I'm right; some six-legged, goggle-eyed life form will be staring back at you.

This pattern continues all over the world, except in the oceans. Insects are the most successful life forms on the planet with well over a million known species – and scientists estimate there could be as many as 10 million! They are also one of the least explored groups of animals. We really understand so little about them and it is this that makes even the most basic backyard entomology exciting.

With more than 400,000,000 insects per acre of normal grassland, it's clear that by looking in the right place at the right time anyone could discover something that has never been known before.

What makes an insect an insect?

What does a delicate, lacy little mayfly have in common with the horned, armour-plated monstrosity that is a Hercules Beetle (*Dynastes hercules*)? At first glance, they are about as similar as a paper aeroplane and a fridge-freezer. But look with scientist's eyes and you'll find many structures that they have in common.

The insect body-plan is highly versatile and can be stretched and moulded by environmental requirements and evolution into many strange and fantastically diverse forms. Adult insects are easily recognised by even the most inexperienced person if they know what to look for.

All insects have six legs, and three main body sections – a head, thorax and abdomen. These may be a little obscured in some cases but if you look hard enough they will be there. Of all the invertebrates only insects have wings, but not all insects have wings! In a nutshell, if it's a creepy-crawly and it is capable of active flight, it is without doubt an insect.

The incredible camouflage of this Leaf Mimic Katydid from Borneo says it all, its form serves its protective function perfectly.

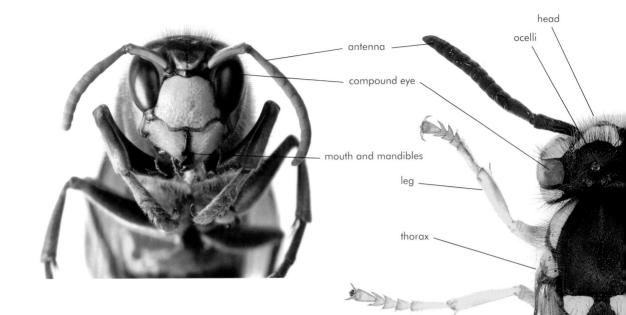

antenna

compound eye

mouth and mandibles

head

ocelli

leg

thorax

foot

2 pairs of wings

abdomen

Head. Sometimes referred to as the head capsule, this houses most of an insect's sensory organs.

Antennae or feelers are moveable organs that do a lot more than feel and touch – they are sensitive to smell, taste and vibration.

Simple single eyes – ocelli – probably just detect whether it is light or dark; compound eyes are much more complex, with thousands of lenses, are very sensitive to movement and give a detailed view of the world. Some insects have both kinds of eyes. Flying insects in particular often have excellent colour vision, and can even see into the realms of colours beyond our own perception, such as the weird world of ultraviolet.

The **mouth** is surrounded by an arrangement of complicated mouthparts. Insect mouths come in a huge variety of different designs depending on what food they process. Some adult insects have no working mouth at all and never eat. Mouthparts can be designed for biting and shearing through plants, leaves and stems, or as meat slicers to mash and carve up flesh. Some have been modified into spades and trowels, to manipulate stuff like mud and paper. Some get stretched to crazy proportions and are used as weapons against prey, predators and even each other – check out the jaws of a Stag Beetle (*Lucanus cervus*). Others are not like jaws at all – for example, butterflies, moths and mosquitoes have sucking tubes. Many insects produce compounds through their mouths. The silk from caterpillars comes from modified salivary glands, while others produce toxins, venoms and digestive juices, which are either vomited or injected through the mouth or jaws.

Thorax. You can think of this as the engine-room of the insect. It is the section with all the legs and wings sticking out of it! Inside are the muscles which drive the legs and wings.

Six legs are useful, providing stability and efficiency; the insect can move in such a way that there are always three legs in contact with the ground. Some legs are modified for jumping, swimming, digging or fighting. Many insects have sensitive legs that can not only feel, but also taste and smell. Crickets and grasshoppers even have an 'ear' just below their front knees!

Insect **feet** have similar sensory devices to the legs and are very sensitive to the environment they are treading on. If you look at an insect's foot under a magnifying lens you'll see it has lots of spikes, bristles and hooks – these are used as a kind of grooming kit in some species. They are used like hairbrushes to keep the antennae, eyes and body clean and tidy. Others are like grappling hooks and crampons, used for grip. Some have a special pad at the tip of each foot, covered in thousands of microscopic hairs with flattened tips. A grease oozes out of glands on the foot, enabling it to stick to surfaces in the same way a wet drinks coaster sticks to a glass. The feet can support more than 10 times the weight of the insect in some cases!

The Giraffe-necked Weevil is endemic to Madagascar. it is one of the most unusual looking insects in the world.

Not every insect has **wings** but when they are present there are two pairs. In beetles and true bugs the forewings are modified as protective coverings for the hindwings. In butterflies the fore- and hindwings are joined by little hooks, so the four wings act as two. Some use them for display, or to produce sound as in the crickets. In the true flies, the rear pair have been modified into little hair-like stumps called halteres – you'll need to look closely to see them at all.

Abdomen. The reproductive and digestive organs, fat storage and heart are found in this part of the insect's body.

An insect's breathing apparatus is a system of crazy internal plumbing – lots of little tubes called trachea that act like the ventilation ducts in a large office building. These open into a series of tiny holes called **spiracles** down the side of the body – up to 10 pairs. These can be opened and closed by the insect to minimise water loss, but the actual 'breathing' is no more than a passive wafting in and out of air through the system. Some insects can be seen to 'pant', with muscular contractions and wriggling of the body to squeeze air in and out quicker than normal.

Butterflies and moths

These insects are well known to everyone, with butterflies in particular the perfect PR agents for insect life as a whole. I started with them as a child and so I'll start with them now. Butterflies seem to flit around whimsically, bringing us the same kind of pleasure as flowers might if they were free-flying. It also helps that they are most active on nice warm, dry, sunny days! Butterflies are not just a set of pretty wings – they face the same challenges in the struggle to survive as all the other 'ugly' insects. Between them the 20,000 or so species have come up with some pretty devious survival strategies; deceit, parasitism, murder, chemical warfare and cannibalism all included.

Moth versus butterfly

Butterflies and moths have wings covered in microscopic **scales**; these give them their famous patterns and bright coloration. They also give the order its scientific name Lepidoptera (lepis = scale and pteron = wing). Close up, they look like tiny overlapping tiles on a roof.

The long tube-like **proboscis** is common to all but the most primitive of moths and comprises two tubes that are 'zipped' up along their length. Each tube has a groove in its surface which when joined together form a central tube, through which liquid food is sucked up.

Generally butterflies rest with their **wings** closed above their body and moths with theirs flat. However, some of the geometrid moths can be seen perching with their wings straight up, and the skipper butterflies hold the forewings up and hindwings flat when active.

Butterflies are supposed to be **colourful**, moths dull and dreary – but find a picture of a Dingy Skipper butterfly (*Erynnis tages*), or look at the pinks and greens of an Elephant

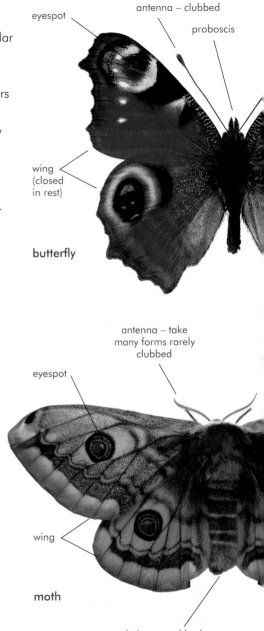

eyespot

antenna – clubbed

proboscis

wing (closed in rest)

butterfly

antenna – take many forms rarely clubbed

eyespot

wing

moth

hair-covered body

Butterfly or moth?

What makes a butterfly a butterfly? Or a moth a moth? Butterflies fly by day, are colourful, have little knobs or clubs on their antennae and rest with their wings held vertically above their body. Moths are exactly the opposite, they fly by night, are dull, have knob-less antennae and rest with their wings flat out. On the whole these rules hold up, but just to confuse things there are some exceptions.

Hawkmoth (*Deilephila elpenor*), and tell me that still holds true. Generally speaking, though, you need to look a bit closer for the beauty of moths, while butterflies are much more in your face.

The very technical names of butterflies and moths, Rhopalocera (butterflies) and Heterocera (moths), describe their **antennae** shape. Rhopalocera means literally 'clubbed antennae' and Heterocera 'different antennae'. However, the interpretation of 'clubbed' is debateable. Some moths, for example the day-flying burnets, look like they have clubbed antennae. But on the whole this rule holds. Butterflies have simple clubs, while moth antennae vary from thin and spindly to bushy comb-like affairs.

Moths usually have fatter-looking **bodies**. Both butterflies and moths have bodies with a covering of hair-like scales; these tend to be a little longer and more pronounced in night-flying moths as they need insulation against cooler night air. These hairs can make the body look bigger. But some moths are slimline, especially the smaller geometrids and the grass moths.

Moths are **nocturnal**, butterflies diurnal. Yet again, this holds in most cases but you'll find exceptions – there are some common day-flying moths. In the tropics there are also a few night-flying butterflies.

Looking like miniature roof tiles, up close (40x) butterfly wings are covered in tiny scales that look like a dusty powder to the naked eye.

The complex feathery antennae of the Atlas Moth have a large surface area to pick up the special perfume called pheromones produced by the female to attract males in the rainforests of Asia.

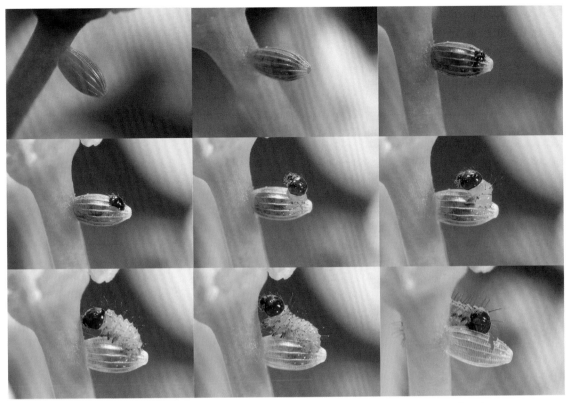

The magic of metamorphosis

Butterflies and moths totally rebuild themselves several times in their lifetime; from egg, to caterpillar, chrysalis and finally to winged adult. The transformation is so extreme that they look like very different animals. This is called 'complete metamorphosis'. Many other insects, including beetles and flies, have the same strategy.

One advantage of this is that adult insects do not compete with their own young for food – they are in effect completely different animals not only in appearance but also in habits and requirements; the winged adults sip nectar and the caterpillars eat leaves.

The egg: little box of tricks

Finding butterfly and moth eggs is a subtle craft. The best way is to watch the adult insects. This is obviously easiest with the day-flying butterflies. A sunny patch of Stinging Nettles is a great place to watch out for egg-laying Small Tortoiseshells (*Aglais urticae*), Peacocks (*Inachis io*) and Red Admirals (*Vanessa atalanta*), while Large and Small Whites or 'cabbage whites' (*Pieris brassicae* and *P. rapae*) can be watched (to the disgust of the gardener) laying eggs on brassicas and nasturtiums.

This sequence of an Orange-tip Caterpillar hatching is just one of the many magical transformations in the life cycle of a butterfly.

Have a look at butterfly eggs through a hand lens and you'll see weird and wonderful designs. Less than 1mm tall, they can look like barrels, skittles, bowling balls, or fine-blown glass, some with designs so intricate they would be more at home among the domes of a Victorian palm house.

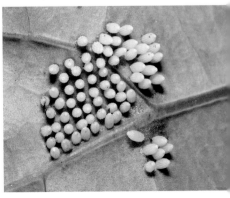

A female butterfly about to lay regularly touches down, tasting with her feet to find the right foodplant. Watch for her to pause and curl her abdomen up underneath the leaf. Keeping a fix on this position, go and investigate and you'll probably find the egg or eggs. Some species are so predictable in their habits that their eggs can be found easily once you have got your eye into their requirements. In Europe the Orange-tip (*Anthocharis cardamines*) is one of these. It has a short flight season and a favourite foodplant, namely Cuckooflower, or Lady's Smock. Carefully scrutinise the flower heads during spring when the adults are on the wing, to find the bright orange eggs.

Some species are easiest to find as eggs. The Brown Hairstreak overwinters in this stage and if you known the sorts of spots likely to be chosen you are more likely to find the eggs than the caterpillars, pupae and even the adult insects!

A Map Butterfly laying eggs on a nettle.

Consuming as a passion: caterpillars

Caterpillars don't have a hard, external skeleton – their bodies are supported by fluid inside, kept at pressure. These soft, sometimes hairy babies may look defenceless, but are spectacularly well-equipped for their two functions in life – eating, and staying alive long enough to proceed to the next stage.

Caterpillars moult as they grow, shedding their skin five times between the egg and the final moult that will change them beyond recognition. First they spin a pad of silk to which they attach their back legs. When they are ready they simply walk out of their skin, which remains attached to the silk, and carry on feeding.

If caterpillars are your chosen quarry, there are many tricks that will lead you to them. You can simply put your naturalist's eyes to the test and go looking for them; but you'll quickly find you're playing a losing game – one they've been playing and winning for some 150 million years!

What isn't there to like about a big juicy Hawkmoth caterpillar? These eating machines are some of the biggest and most spectacular caterpillars a naturalist can discover.

Search at night – many caterpillars are nocturnal, active only when the eyes of most predators are shut. Look for feeding evidence such as chunks and crescent shapes taken out of the edges of leaves. Large caterpillars such as those of hawkmoths are often given away by distinctive-shaped droppings scattered under suitable food trees or overhanging hedges.

They use lots of tricks, colours and patterns to make them nearly impossible to spot by day. But at night they relax a bit and start moving around, often heading for the more succulent leaves at the ends of twigs and branches. Using a torch you can pick them out rather easily. Searching by artificial light disrupts their counter-shading. Torch beams also effectively narrow your field of view, forcing your eyes to only concentrate on that which is illuminated.

When attacked by a predator a Puss Moth caterpillar puts on quite a display, pumping up its body, thrashing from side to side and whipping out two red thread-like tails.

Hairy horrors

Caterpillars can be rather fragile and delicate, but it's worth bearing in mind that they are as good at staying alive as they are at eating. Some of the hairy and bristly ones are far from cuddly, but are engaged in out-and-out chemical warfare. Armed with nasty poisons, they bristle with spines and toxin-laced hairs that will at best make your skin crawl and at worst put you in hospital. There are some tropical and American species, such as the flannel moths, that are particularly nasty.

I once inadvertently shared a sleeping bag in the Australian tropics with what had been a hairy caterpillar (it was bald by the time I discovered it) that made me feel like I was being burned alive! I removed the insect and spent the rest of my night in agony. Only in the morning did some relief come, in the form of half a box of antihistamine tablets, some aloe vera cream, and the humiliating business of standing in a forest clearing half-naked while my companions got the caterpillar hairs off me using strips of sticky tape. The moral of this tale is that caterpillars, especially hairy ones, are not to be touched with bare skin!

Many caterpillars warn you of their painful potential by being outlandishly coloured. Those that live gregariously sometimes go in for joint predator evasion. Clusters of Peacock and Small Tortoiseshell caterpillars will at first all twitch in unison, and then curl backwards and ooze a green gob of foul liquid at you, and if this fails will promptly roll off the leaf and fall down to the base of the nettles on which they feed, where they are extremely tricky to find. If you are collecting, use a spoon underneath them to catch them before they escape. Others, especially those of moths that roll up leaves for protection and make little fortresses in their food, can move surprisingly fast and catch you out when you unroll their homes. Some even thrash their bodies from side to side in a fit of wriggling, which makes them impossible to pick up.

To find a beauty like this Sycamore Moth caterpillar is every naturalists dream – but don't get carried away and stroke it.

Bashing the bush

With this method, you shake caterpillars out of the bushes or trees in which they live, using a stout stick or broom handle. You'll also need collecting pots, pooter and something to catch the caterpillars on. You can purchase purpose-made beating trays, but an upturned umbrella or bed sheet will do. Hold your umbrella or spread your sheet out below the bush to be beaten and then shake the foliage. One sharp tap and insects should lose their grip and fall. Be careful not to damage the trees themselves.

In summer you can get many species from just one type of tree or shrub and you'll get to meet a selection of species, from hairy toothbrush-like kinds to simple green jobs. You may be surprised to see a couple of sticks get up and loop off – these are the inchworms or loopers; caterpillars of the geometrid moths. If you can't identify them with a field guide you can rear them up to the adult insects – more about that later.

The perfect stick mimic, the caterpillar of a geometrid moth is also the source of the name 'inchworm' and their name which means 'earth measurer'.

Caterpillars of browns and skipper butterflies can be found as they crawl from their hideouts at the bases of grass blades. Get down on the level of the grass and shine your torch to pick out the shapes of caterpillars as they feed.

The third act – a perfect pupa

Fully mature caterpillars finally stop eating. They change colour and start charging around looking for the right place to turn into a pupa or chrysalis. It is at this time when people find them roaming across patios and pathways. Some pupate on or in the ground, others dangle from a stick or branch. Some nymphalid butterflies hang upside down, curling in a 'J' shape, while the 'whites' remain head up, but spin a little waist harness around their middle. Once secured, the caterpillar's head capsule pops off and the skin is shrugged off for the last time, revealing the pupa or chrysalis skin beneath.

What goes on inside is an amazing shape-shifting, from caterpillar to the adult insect. It was once thought that the caterpillar dissolved

You can see how the name 'chrysalis' came about when you look at this Painted Lady chrysalis – the mirrored surface play games with the light and makes it difficult to see in nature.

into a living soup, and then rebuilt itself into a adult insect. Thanks to a scientific process called micro CT scanning we now know this isn't completely true. The gut and breathing system stays in place but is modified. The muscles and other tissues break down, but in an ordered way, forming clumps of cells that reposition themselves – you can think of it as a chunky soup rather than a thin one.

The word pupa means 'doll' or 'puppet', and describes this stage of moths and butterflies (and beetles, bees and flies). Chrysalis means 'golden', and is used exclusively for butterfly pupae, referring to the metallic sheen on some.

When the pupa splits, after between two and 24 weeks depending on the species, the adult emerges and at first sight is a bit of a disappointment – a scrunched-up creature that is all eyes, antennae and six ungainly legs that look ill-designed for walking. With time this chimera drags itself somewhere it can hang freely. It pumps blood into wing veins, they expand and the ugly duckling becomes a beautiful, elegant butterfly (or moth).

Pupae have an array of extraordinary ways of hiding from hungry predators' eyes. Some chrysalises have sculptural qualities, bedecked with weird appendages, lumps, bumps and flanges all designed to camouflage. Some play tricks with the light by incorporating mirrored panels into themselves to reflect their background.

Looking like a doll, this silk moth pupa displays the blueprint of a future moth – you can make out the wings, legs, antennae and eyes.

Digging in the dirt

Death's Head Hawkmoth pupa

Many species of moths make their loose cocoons and pupae below the surface of the soil. To find these subterranean sarcophagi you need to dig. Take an old margarine tub of soil/moss, a trowel and a hand fork and gently dig around the bases of trees, up to 10cm (4in) down. With luck you'll find these entomological treasures glowing at you, looking like miniature bronzes in the dull earth.

In my experience in Europe certain trees are more productive than others; willows, oaks and hawthorns are best. It is very hard to identify a moth species from its pupa alone, so if you are curious as to what you have you need to hatch them out. If you have found your pupae in autumn or winter they will probably not hatch until the following spring, but any other time of the year it's anyone guess and you'll just have to keep checking; they could hatch at any time.

Moths tend to rely more on a cocoon surrounding the pupa for protection. They can incorporate the defensive hairs of the caterpillar or bits of the surrounding habitat to form a camouflage or tough bag or box; the Puss Moth (*Cerura vinula*) constructs a fortress of chewed wood particles and silk that you need a chisel to open up! Most, however, are just well hidden and are formed in loose cocoons, out of sight beneath the soil.

Like a solid piece of wood, the Puss Moth chews up wood, mixes it with silk and saliva and produces a cocoon so tough it can only be opened with force.

The crescendo – the adult emerges

The moment that the pupal skin splits is something that every naturalist should endeavour to see. The only real way of witnessing the whole act, from the first splits in the seams to the fully formed adult's virgin flight, is to witness a captive animal emerging. You can usually tell when this process is imminent, as the colours of the wings show through. Be patient and keep checking back – with luck you will be rewarded.

The sole purpose of an adult butterfly or moth is to breed and much of what they get up to on the wing is centred around finding a mate and laying eggs. Some moths don't have functional mouthparts, and are fuelled by fat reserves built up as a caterpillar. Others need breaks to feed and refuel on nectar and other nutrients, a habit that you can use to your advantage.

Butterfly-watching through binoculars is a great way to witness their behaviour without disturbing them and for me is a much more relevant and pleasurable pursuit than collecting them. There are also many great field guides available to help you identify and interpret the various behaviours you witness.

Everyone need to see the everyday miracle of a butterfly emerging from a chrysalis – it takes my breath away every time. This is a Painted Lady butterfly.

Bringing up butterflies (and moths)

For a really intimate look at the life phases of a butterfly (or moth), you have to grow your own. Collect eggs or caterpillars from your local patch – that way you will know what plant they feed on, and have a ready supply of it close at hand – or buy them from a specialist supplier. Collecting at the egg stage reduces the risk of sneaky little parasitic flies and wasps getting there before you. It's quite distressing when you have invested a lot of time, effort and energy into rearing caterpillars only to have them burst open with small maggot-like creatures one day!

Large White butterfly eggs, which will hatch in a week or two.

Eggs are best kept in small, airtight plastic boxes to protect them from drying out or being scattered around the room by a misplaced sneeze! There's plenty of air in there, but open the box every other day to ventilate the eggs and breathe on them gently to keep them moist. Keep the box out of direct sunlight and check it every morning – caterpillars often hatch at night.

Caterpillars are tiny when newly hatched – as little as 2mm (½in) long. Many consume their own eggshells so don't be too keen clear up until your caterpillars have wandered away in search of other food. Transfer them to a slightly bigger plastic box using a spoon and a fine paintbrush. The box should be lined with tissue paper and provided with a leaf or two of foodplant. Clean the box every day and keep the caterpillars well supplied with fresh leaves that are neither very young (which will upset their stomachs) nor old and leathery (hard to chew). Caterpillars that are feeding happily will start to produce dusty droppings known as frass, which will remind you to keep changing the tissue paper. Caterpillars grow fast, and soon need to be moved somewhere more spacious with better ventilation. Anything from a larger margarine tub to a shoebox makes a good home at this stage. Provide cut stems of their foodplant daily in a small container of water (old film canisters are ideal) and wrap tissue paper or Plasticine round the stems to stop the caterpillars falling in and drowning. Clean out daily and upgrade the accommodation when it looks too crowded.

Simply cut a hole in the lid, then glue or staple a fine mesh, gauze, or muslin over this, making sure the holes are not big enough to let the caterpillars through.

Pupae After five moults, caterpillars are ready to pupate. Provide butterfly caterpillars with sticks or twigs from which the pupae can hang. Moth caterpillars will either cocoon themselves in eggboxes or bury themselves in soil. Be patient, move the pupae to an airy mesh cage that will be suitable for the adults when they appear, and

Cuts stems of the foodplant can be placed in a small bottle or jar of water. To prevent the caterpillars from drowning stuff tissue paper or Plasticine in the gap around the stem.

provide branches or netting for them to climb up. Spray the pupae with a mist-sprayer daily.

Emergence The sign something's about to happen is when the pupa becomes semi-transparent. Check every few hours, especially first thing in the morning. Once emergence has happened, your successfully reared butterflies and moths can be released back into the wild. Alternatively, if you have a pair of moths, you might like to see if you can complete the cycle by getting them to mate. Species without mouthparts, such as Poplar Hawkmoth (*Laothoe populi*), are easy as you don't have to feed them, but all species will need some plant material to lay eggs on. Sex them by antennae shape (often feathery in males), and also abdomen size – females are fat and chunky, males are thinner. The other magic ingredient is a bit of peace and quiet (moths are very sensitive to vibration and disturbance). So place them in a cool dark room and put a 'do not disturb' notice on the door. With a bit of luck your moths will mate within a few days, the female will lay eggs and you can start all over again.

As they grow, continue to clean and refresh the food on a daily basis and if necessary upgrade to a new box.

Making a moth mix

Butterflies spend their days seeking sugary fuel for their activities. Moths do the same thing at night, and locate food using their sensitive sense of smell. They are attracted to scented flowers, fermenting fruit and oozing sap. You can utilise this behaviour by using a technique known as 'sugaring'.

You'll need a big pan, sugar, molasses, fruit juice, dark rum and understanding family members, because you will fill the house with a very sickly smell. The whole mess should be gently heated and stirred adding water or sugar to get the consistency just right. It must be runny enough to paint onto a fence post or tree trunk easily, but not so runny that it dribbles away. At dusk on a warm summer night, take a jam jar-full and with a stout paintbrush apply the mix generously to tree trunks and fence posts. Return an hour or so later with a torch and see what you have lured out of the night.

This Old Lady moth is one of the species that rarely comes to light but finds a sugary sticky mix irresistible.

Moths and the 'electric flame'

Since early man first banged some flints together to produce fire, it has been well known that moths and other nocturnal insects are attracted to light, whether a flame or its 20th-century equivalent the lightbulb. One theory as to why they do this concerns navigation. If a moth is flying from A to B and keeps the moon on the same part of its eyes, it is going to go more or less in a straight line. But if you shine a light source such as a lightbulb close to the moth and it mistakes this for the moon, if it tries the same thing it will fly around in circles and get confused. Moth traps exploit this behaviour.

The best moth traps tend to use special lightbulbs such as mercury vapour bulbs that are highly attractive to moths and other nocturnal insects, giving out a high degree of light in the ultraviolet part of the spectrum. They are used by professional entomologists the world over, and if you are serious no other light source will do.

The simplest trap is a white sheet, suspended on a string or frame between trees. The light source can be a strip light suspended at the top of the sheet, or a regular bulb mounted on a post in the centre. This is easy to transport, and a great way to demonstrate the technique and the insects to a group of people, but is difficult to manage on windy nights. There is no actual trapping device either, so you have to man the sheet constantly (most moth activity occurs just after dark, with other peaks at midnight and before dawn).

There is nothing like an evening around a mercury vapour lamp to shed a little light on those mysteries of the night – the moths. It can be quite a sociable learning activity too.

When a sheet is set up in a rainforest at the end or beginning of the rainy season, it can deliver one of the most exciting spectacles an entomologist could ever hope to see. I've seen sheets completely covered with insects, and anyone who stands in front of the light also gets smothered by insect life.

The other popular designs are true traps, useful if you are taking part in a study where you are to be trapping regularly, and/or you don't wish to man the traps all night. They comprise a light source, and a collection vessel into which the insects are funnelled. This is usually furnished with egg cartons to give the moths somewhere dark to hide, reducing the chances of them flapping around in a frenzy and damaging themselves. The other consideration is of course a power supply. A long flex helps position the trap a good distance from other competing light sources. When trapping a long way from my own home, I have found complete strangers willing to share their electricity for the night in exchange for a morning lesson about their moths. I even befriended the entire staff of an alpine monastery in Switzerland when running a light trap for migrant moths flying through the passes.

The other alternative to mains electricity is to take a small portable generator with you and run the trap off that. **In any case, the nature of this activity involves running electricity outside, with the inherent risk of moisture – the combination of these two can kill. So be sure, be careful and always use a circuit breaker.**

For spontaneous entomology, any location where there is a light on all night is worth checking out. Many of my own entomological highlights have been in unexpected but lit-up places. A Giant Atlas Moth (*Attacus atlas*) in a Malaysian garage forecourt plunged everything into a total eclipse as it spun in ever decreasing orbits around the single lightbulb. A lantern bug (spectacular bugs of the neotropics) was in a South American shop window, and a whirlwind of dozens of spiralling Stag Beetles caught my attention in the halo of a street lamp in south London. In Africa I was joined in my pursuit by a hyena that was hoovering up the unfortunate insects before I could get to them, and my very first living Convolvulus Hawkmoth (*Agrius convolvuli*) came from the gents' toilets in an M4 service station!

A Common Quaker moth in flight.

~ MAKING A MOTH TRAP ~

Not all entomologists are insomniacs, so constructing a simple trap means you can go to bed and let the moths get on with it. Most of the process is creating a collecting vessel in which the living insects can be safely trapped until the morning. The electrics really need a bit of expertise – no moth is worth electrocuting yourself over, but with the rising popularity of moth trapping in recent years there are many commercial models available and even if you don't want to spend too much money, you can invest in a simple kit form for the electrics, which is well worth it. Here are instructions for a simple one; simply substitute the electrics mentioned here for a kit form.

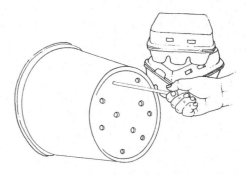

Step 1

You will need:

- plastic bucket
- candle
- meat skewer
- eggboxes
- 2 sheets of clear plastic, one approximately 70cm (27½in) square (whatever size you need to make a funnel that fits into the top of the bucket), one about 20cm (8in) square
- strong glue
- scissors
- 3 strips of wood 2cm (¾in) square and about 30cm (12in) taller than the bucket
- piece of wood 10cm (4in) square and 1cm (½in) thick
- saw or file
- lightbulb and bulb mount – a bulb that gives off an element of the ultraviolet spectrum is best
- long flex
- cable clips
- plug
- cloth tape

Step 2

1. Warm the tip of the skewer in the candle flame and use it to poke a few drainage holes in the bottom of the bucket. Put the eggboxes in the bucket.

2. Make the large sheet of plastic into a funnel and sit it in the bucket, using the glue and scissors to hold it together and trim it into shape – it should stick out of the top of the bucket by about 2cm (¾in).

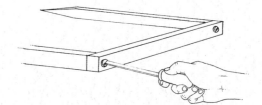

Step 3

3. Screw the strips of wood together to make a shape like football goalposts. Use the saw or file to sharpen the ends that are to go in the ground. The crossbar needs to be 20cm (8in) above the bucket once the posts are pushed into the ground.

4. Set up your light fitting by wiring the flex to the bulb and screwing the mount to the smaller flat piece of wood. Attach this to the underside of the crossbar and tidy up the flex with cable clips.

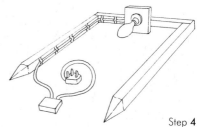

Step **4**

5. Screw the smaller piece of plastic to the top of the structure to keep the rain off.

6. Choose a position on soft ground so that you can shove your 'goalposts' into the earth, and make sure it is near enough to a power point that you can plug the whole thing in. (Always install a circuit breaker for safety.) Position the bucket and funnel, then press the goalposts firmly into the ground. Switch the light on at dusk and go to bed when you feel like it.

7. In the morning, turn the light off, unplug the electricity and carefully remove the bucket. Take it to a shady place so as not to dazzle your captives and look in the eggboxes to see what you have found.

Step **5**

Step **6**

Step **7**

Put them on ice

If it is a warm morning, your moths will be beginning to wake from their slumbers. They will beat themselves into a dusty pulp if they can't escape. If you wish to make a systematic and thorough record of the night's takings, use a cooler or fridge. Not too cold – just enough to slow down the insects' activities. Cooling insects for short periods of time is very useful for identification and photography.

Calling moths

Females of certain lepidopteran species lure their mates from far and wide, using the power of perfume. This behaviour can be used to great effect by entomologists in a technique called 'assembling'. A traditional way to obtain a mate for a freshly emerged captive female, it is very much a dying art, but a spectacular demonstration of the phenomenon.

The first time for me was when I was invited by an entomologist and breeder of Emperor Moths (*Saturnia pavonia*) to join him, and a virgin female moth in a net cage, on a heath in southern England. The weather was warm and sunny with a light breeze. She was a big, plump specimen, her abdomen swollen with hundreds of unlaid eggs. The abdomen tip displayed a small glossy gland, which if I was a male moth would be heaven-scent, as it emanated a strong pheromone. This is a species-specific smell carried up to a mile downwind of the 'calling' female moth.

Within 20 minutes the effectiveness of this was clear. The males came zigzagging low over the heather towards the cage, and soon the female inside it could hardly

A female Emperor Moth casts her perfume to the breeze and attracts dozens of suitors.

be seen for the flapping and twitching of about 40 expectant and hopeful males! They had found her by smell alone, collecting and analysing scent particles on their plumed and feathery antennae (incidentally, this is a way to sex adult moths that use pheromones; males always have a much larger surface area to their antennae, and their antennae have a more obvious comb-like appearance than the females' do). Many species of moths, butterflies and other insects utilise pheromones in this way, especially those with flightless females.

Butterfly-baiting

Many insects have requirements that we humans can satisfy. The most obvious example is planting food supplies for the adult insects, in the form of nectar-rich plants. Brambles, mints and other herbs always seem to have plenty of winged things buzzing around them.

However, some of these beautiful and heavenly creatures have remarkably unsavoury tastes! Travel a tropical road in Africa and the highest concentrations of butterflies will be competing with flies for the goodies oozing from a patch of urine-soaked earth or a steaming pile of dung! I have seen this to such extremes that an elephant dropping resembled a giant multi-coloured football. Naturalists use this to their advantage to lure in their desired quarry. Various baits can be used from overripe brown and soggy bananas, to animal excrement and rotting corpses. All are sources of sugars, minerals or salts, or just moisture that may be in short supply elsewhere.

If you don't fancy leaving dung in your garden, try placing a shallow tray of mud and water near your butterfly garden. You'll probably attract butterflies, and bees and wasps will also stop by for a drink. You may even get a potter wasp collecting materials for its nest.

The caterpillars and larval stages of other insects can also be catered for in the garden. Nettle patches in a sunny corner of the yard may attract the big showy nymphalid butterflies to breed (don't let them get too tall, I cut mine in rotation so there is always lush fresh growth). Other foodplants such as Hedge Mustard, Honesty and various wild members of the cabbage family also work well. Do your research, make your observations and plant accordingly, and you'll make a lot of caterpillars and other insects very happy.

These Small Tortoiseshell butterflies are attracted to the late flowering sedum flowers, giving them a welcome nectar meal before they prepare for hibernation.

Dragons and damsels

The dragonflies and damselflies are great ambassadors for the insect world, second only to the butterflies. Although you can come across them far from fresh water (I once saw one flying out at sea, more than 100 miles from the nearest land) the best place to catch up with these brightly coloured shards of insect life is by a body of water on a hot, sunny day, and pretty soon you'll appreciate just why I consider these insects the soul of a summer's day as they dash, dart, whizz and skip around. Their order is called Odonata, which means 'toothed jaws', and this gives away their game a bit. They are the assassins of the sky, trawling the air for other small insects. Even the tiniest damselfly is a ruthless murderer of midges – some consume up to 20 per cent of their own bodyweight a day. As aquatic nymphs they are just as deadly to other small denizens of the ponds, steams and rivers they inhabit.

This Twelve-spotted Skimmer demonstrates rather effectively just why these insects capture our affections. Notice the way it holds its wings flat out to the sides and has a big chunky look characteristic of a dragonfly.

Damsel or dragon?

This is quite easy once you know what you're looking for. Dragonflies are bigger, with thicker, chunkier bodies, and have a more direct and purposeful flight. Damselflies are slim and fluttery. They 'hold' themselves differently when perched; dragonflies rest with their wings held on or near the horizontal plane, flat out like the wings of an aircraft. Most damselflies rest with their wings together, above their back. However, when dragonflies have just emerged from their nymphal skin and their wings are drying, they hold their wings closed above their back.

The more delicate damselfly in a typical resting posture of these often smaller and more dainty cousins of the dragonflies.

Dragonfly nymph

Only damselfly nymphs have the three leaf-like tails (actually gills) making them very distinctive and completely different to the dragonflies who keep their gills in a chamber in their abdomen.

The nymphs are highly variable in colour, shape and size, but dragons and damselflies look completely different and it's all down to where they keep their gills. Damselfly nymphs are delicate and slender like their parents, and have three leaf-like gills at the tips of their abdomens. Beware confusion with mayfly nymphs, which also have three 'tails', though these are not flattened. Dragonfly nymphs are generally more robust-looking and, while they may have a few spiky projections at the tip of their abdomen, their gills are internal.

Dipping for dragons

A large dragonfly can reach top speeds of around about 36kph (22mph) (even damselflies can reach 10kph (6.2mph)). So it makes sense to start with the slowest stage of their lives, which means getting to know the nymphs. They are aquatic, very subtle, slow and well-camouflaged beasts that prowl or lie in ambush for smaller prey. You'll need a dip net to hunt them down. You can turn them up from among aquatic vegetation, mud, sand and silt of the bottom of a pond.

Damselfly nymph

For species that live in flowing waters, kick-sampling is one way of catching them. You kick around in the loose substrate of the stream bed, with your net positioned downstream to collect anything that has been disturbed and set adrift. Turn out the contents of your net into a white tray and be patient. Even if you think you have captured nothing but sludge and weed, you'll be surprised by what will wobble and walk about after a few minutes of time to settle down. You'll need a selection of clear pots or even a tank to appreciate what monsters these young insects really are.

The original jet

If you find the elongated, robust forms of hawker dragonfly nymphs, you are in the presence of nature's original jet engine! The gills of these insects are inside the abdomen and water is flushed over them by actively pumping it in and out.

These same muscles can be turned to a rather surprising use. By rapidly squirting the water out backwards, the nymph puts Newton's third law of motion to good effect and is catapulted forwards! This can be nicely demonstrated if you gently lift one to the surface of the water. It will try and escape, frantically deploying its jet; angle the insect's abdomen tip upwards and you'll see it squirt this jet of water up and through the surface, like a living water pistol.

Try and catch this Southern Hawker and it'll surprise you by jetting off – squirting water rapidly out of its backside.

Face off

How does the sluggish-looking nymph manage to catch active insects, even small fish and tadpoles? Its secret weapon is a set of extendible jaws, the labium, which at rest is folded up under the head. When the nymph spies a passing meal, it fires this hydraulically powered arm-like lip forward and stabs, grabs or spears the prey.

In some species the labium is tipped with a pair of sharp spines. A few have one that looks and works like an industrial grab-and-mangling machine. If you watch a nymph in a pot or tank with potential prey present you may see this dastardly gadget in action – well you think you do. The speed of this feeding reflex is so fast (16–25 milliseconds) it's actually impossible to follow with the eye.

A dragonfly jaws are formidable enough without the fact that they can shoot out from the font of its face to capture prey.

Collecting exuviae

I like to collect stuff, it's how I learn. If you are a bit like me you might want to collect the empty nymphal cases of dragonflies and damselflies. These husks are called exuviae and are perfect hollow representations of the creatures that once lived in them, eyes, jaws and all.

Look for them are around the edges of ponds and streams; focus your attention on stones, roots, tree trunks and any emergent

1

2

3

4

vegetation (that's plants such as reeds and rushes that stick out of the water). What you're looking for are small empty, brown husk-like skins clinging to the surfaces. It is to here that the nymph would have crawled, having finally divorced the aquatic world before the adult insect came busting out in a moulting process called ecdysis.

Collecting them is often simply a case of picking them up, being careful to unhook their delicate legs. But sometimes they can be out of reach, and there is nothing more frustrating than seeing a lovely-looking specimen attached to a reed, in deep water or mud that you simply cannot reach. To retrieve it you can use a yogurt pot attached to the end of a pole – all you do is scrape the edge of the pot upwards and the exuvia should drop into the pot (this can be tricky in windy weather!).

You can keep them indefinitely once fully dried out, in labelled specimen pots/tubes on a shelf, or they can be pinned or glue to dark card, which really shows them off a treat. If you are a real perfectionist you can even arrange them in life-like poses or with their 'killer' labium fully extended. This is best done by popping the exuvia in a pot of hot water to soften it. Carefully remove it after it has softened, pin it to a cork or polystyrene tile through its middle, then using tweezers and cocktail sticks carefully arrange the legs and the 'mask'. Hold them in place by crossing pins over each other. Then leave it to dry in a warm and airy place until it has gone crispy and

The old empty nymphal skins are a natural treasure, beautiful in their own way, each species can be told apart and the skins can be collected and arranged easily.

1. If the exuvia is dry and crispy, first soak it.

2. Then using forceps transfer it to a pinning board.

3. While it's still supple – using pins arrange the limbs etc.

4. Leave to dry out and you'll have a perfectly mounted specimen.

set again. Remove the pins, and glue it or pin it to a card or tile, then label with date and location.

You can identify the species from their exuviae alone (there are excellent field guides available for just this).

Where to hunt dragons

First of all you need to get yourself to some water. Dragonflies and damselflies can turn up anywhere, especially newly emerged ones which take this opportunity to spread their wings and look for new homes.

The avid dragon- and damsel-watcher needs a few bits of equipment. Close-focusing binoculars are handy, because a lot of the action does take place in the middle of the pond or river. A butterfly net, not a pond net (pond nets are too heavy and can damage frail insects such as damselflies) can be handy for getting a close look at these insects and separating some of the similar species, but handling them without hurting them requires skill and patience and is really for those patient and expert at handling insects. A good field guide is needed if you want to put names to faces.

Dragonfly-watching is easily as fascinating as bird-watching. Just get yourself comfortable on a still, sunny summer's day, sit back and watch. You'll see males patrolling territories, and dog fights as they buzz each other and chase rivals away from their little patch of the sky. You'll witness the hunt, with some species continuously hawking after other insects, while others perch and sit, wait and pounce. You will almost certainly witness some breeding behaviour too.

Mating damsels are often seen flying in tandem, the male towing the female around, gripping her behind her neck with his claspers. The full-on act itself is in the position known as 'the wheel' with the female curving up and around to join the male at the very base of his abdomen. Following on from this is egg-laying in or by the water, which may involve precision placement (as in hawker dragonflies and most damselflies), or a more scattershot approach.

Anywhere near water is a good place to start your dragon quest.

Damselflies in the characteristic mating position called the wheel. The male is the bright blue insect.

~ REAR YOUR OWN ~

As with caterpillars, you can rear dragonflies and damselflies from nymph to adult. Choose a species from a 'still water' habitat as these are easier to keep. Set up a fish tank as for any other aquatic life (outside is best so that if you miss the moment the insects will be free to fly away). Don't put a lid on it; instead provide a selection of twigs, branches or reeds that stick out of the top for the emerging dragons to climb up.

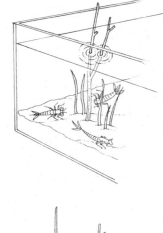

In spring or early summer collect two or three large nymphs for a tank about 60cm (24in) long. Choose the biggest nymphs, with well-developed wing buds, as these are more likely to metamorphose during the coming season. If you choose small ones you may well be in for a long wait of several years!

Feed your nymphs on small worms from the garden. Just prior to emerging they'll hang around near the surface as their gills stop working and they need to start breathing through the spiracles in their thorax. It is now that you should keep an eye open for the grand finale! It usually happens in the early morning or evening, so to increase your luck, step up your vigil at these times of the day.

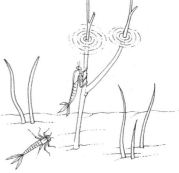

If you are lucky you may witness one of the most beautiful sights in the insect world. Following its slow crawl up a stem, the nymph splits open along a natural weak spot at the back of its head and a brand new dragonfly is born from the brown monster. If you miss the emergence be patient and try again; it took me many attempts before I finally got it right, although you'll have a consolation prize in the form of a very fresh exuvia to play with!

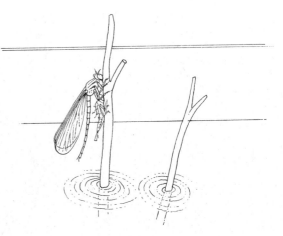

Fiddlers of field and forest

From the fields and meadows of temperate regions to the steamy tropical forests, there aren't many habitats that don't hum, scratch and buzz to the tune of these insects. As a group they are very diverse in appearance although most are well camouflaged. Some fly, others don't.

Most make themselves known to us by sound, as they are often well hidden in the undergrowth or thick grass. Some have an uncanny ability to throw their 'voices' – even when you think you should be getting close, the chirping always seems to be coming from the next clump of grass! There are also those that live underground, such as mole crickets, or in caves and cervices. There are some 22,000 species in the world.

It's just not cricket: telling them apart

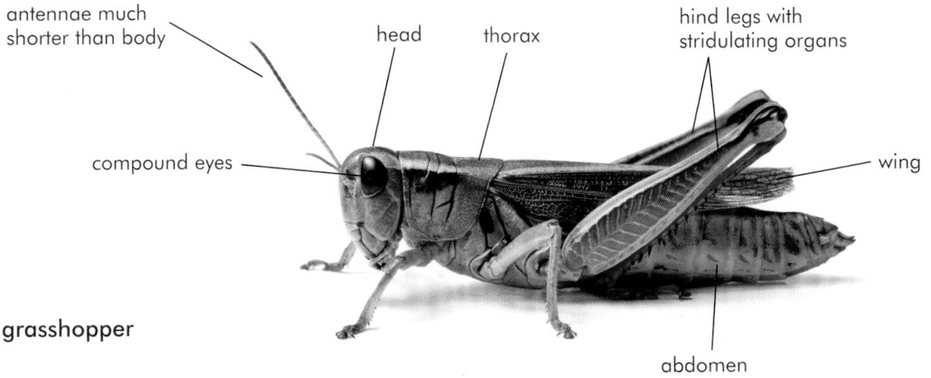

antennae much shorter than body

head

thorax

hind legs with stridulating organs

compound eyes

wing

grasshopper

abdomen

antennae larger than the body

head

thorax

hind legs

wings with stridulating organs in the male

compound eyes

front leg has an 'ear' or tympanal organ

abdomen

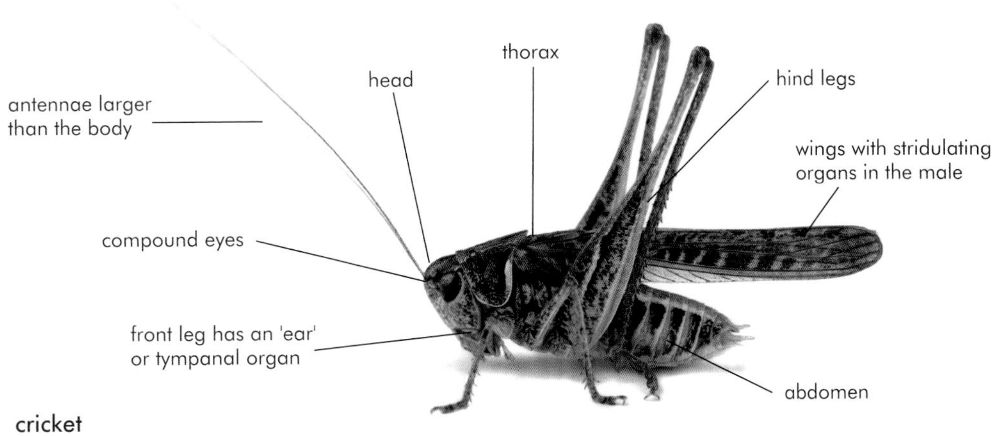

cricket

As a group these beasts are relatively familiar and easy to recognise. While we may all know what they look like, scientists are still trying to agree on what to call them. Some refer to their order as Orthoptera, which means 'straight-winged', but crickets and grasshoppers are best known for their enlarged rear pair of legs, used for jumping. That's why some prefer the name Saltatoria which means 'to jump' in Greek.

Checking the **antennae** is the easiest and quickest way to separate the two groups. Most crickets have whip-like antennae, longer than their body, but grasshoppers' antennae are much shorter and thicker.

Orthopterans' 'ears' are called **tympanal organs**. In crickets they are found on the front pair of legs – if you look just below the 'knee' you'll see a pair of holes. Grasshoppers have theirs hidden at the base and sides of their abdomen, and they are tricky to see as they are often hidden by the wings.

The last of the three pairs of **legs** are larger and contain strong muscles that generate the jumping force. In some species they are armed with lots of spines and spikes for defence and feeding.

They have very effective compound **eyes**, as anyone who has ever tried to catch one will attest. Some also have simple eyes (ocelli).

The **mouthparts** comprise main jaws for mashing up plants and animals, and palps either side for feeling and tasting the food. A thin pair of pleated and folded frail **hindwings** is hidden behind a pair of protective leathery **forewings**.

Crickets have 'ears' called tympanal organs, which are found on the front pair of legs. If you look just below the 'knee', you'll see a pair of holes.

Stridulating organs are the bits that make the noise. Grasshoppers 'fiddle' using their hind legs as a 'bow', rubbing a set of pegs on the inside of the thick bit of the leg against a raised vein on the forewing, a bit like running your nail down the teeth of a comb. Crickets rub a serrated rib of one wing against the rib of the other. On most crickets' wings you can see a clear, rounded area called a 'mirror', which is used to amplify the sound.

The prominent **ovipositor** (egg-laying tube) in female crickets gives the cricket suborder its name Ensifera, from the Greek meaning 'sword-bearing'. You can see this even in nymphs, and it is a good way of sexing crickets. Grasshoppers (suborder Caelifera, meaning 'chisel-bearing') have smaller and less well-defined appendages and it's a lot harder to tell male from female.

Living thermometers

Crickets' activity can be measured by how much they 'sing'. When the air temperature is around 30°C (86°F) they are fully charged. Just listen to a field at the height of summer; the place is a cacophony of sound with crickets and grasshoppers scratching, chirping and clicking all over the place. The same meadow on a dull overcast day will generate a completely different soundscape. Great Green Bush Crickets (*Tettigonia viridissima*) (below) have one of the loudest songs of any European insect and when it's very hot they will sing an almost continuous trill, but as the air temperature drops down to less than

15°C (59°F) the individual chirps can be detected. You can test this by using the chirps of these and other bush crickets to tell the temperature. Using a watch, count the number of chirps in 15 seconds; do this several times to get an average. To find the air temperature in Celsius, divide this number by 2, then add 6. Your answer should be the temperature in Celsius within a couple of degrees. Double-check with a thermometer. How accurate is your species?

The buzz in the bushes

While their technique for producing these sounds may be different, the reason crickets and grasshoppers sing isn't dissimilar to that of singing birds. They are communicating with one another in a habitat that is too dense to see each other in. It is usually the males that make most of the noise, singing to attract a mate, and keeping other rival males at a distance. The females make quiet encouraging little chirrups to tell a male he is on the right track.

Diurnal cricket male on the left 'singing' with raised wings as he courts the female on the right.

Like birds each species can be identified by its calls. Some have more than one sound – a 'serenade' for the females, a 'war cry' reserved for rival males, and a theme for fighting and keeping territory. Most of these songs are well within the audible range for humans, but with some, such as coneheads (a genus of bush crickets), people aged over 20 may not be able to hear the song in its entirety.

Cricket season: finding and catching them

Searching for the animals that are so boldly creating their music all around you can be very frustrating, especially as they do very passable impressions of inanimate objects, from plant leaves and stems to rocks and stones!

Just when you think you are getting close to your target you realise it's throwing its 'voice'. Or even worse, despite your stealthy approach it detects you, stops singing and then you hear the 'flick' as the insect launches itself even further from your grasp with a deft contraction of its saltatorial legs! To study these animals we need to overcome all these neat defence strategies.

As you might expect, grasshoppers live in grassy field edges, meadows and the like and you can certainly find males by their song, but you'll miss the females and young nymphs. Using a sweep net is a great way to sample all. Sweep the net through long grass or other non-thorny vegetation a couple of times and then investigate the bag – it normally isn't too long before you find a grasshopper or two. Bush crickets prefer hanging out in bushes and thicker vegetation. For these you can use the **beating tray** technique used on page 186.

Look in the bushes for Bush Crickets – with their long antennae to help them most crickets live in tangled three-dimensional habitats, although finding them can prove as much of a challenge as catching them.

Longer, thicker, thorny vegetation such as nettle and bramble patches or even hedges and shrubs provide other challenges, and often you'll know a beast is in there but there appears to be no way to reach it. However, it is worth walking up or sweeping the areas adjacent to the songster, as more often than not other individuals will be in the area.

Triangulation tactic

Here's a tip for when there's a group of you and you want to find a noise-making subject (this works just as well for frogs, birds and cicadas as it does for crickets and grasshoppers). Spread out as much as you can in the vicinity of the 'singing' animal, moving slowly and carefully so as not to disturb it. Then all point to the spot you think the sound is coming from (sometimes making your ears more directional by cupping your hands behind them can help). The spot where all your points cross is where the animal is. Then all move slowly towards it, closing in on the point of interest. If you're lucky and stealthy you'll often be able to pick out the subtle twitching of the animal as it broadcasts to the airwaves.

Circle of life: incomplete metamorphosis

Grasshoppers and crickets go in for something known as incomplete metamorphosis. They start life as an egg, then go through a series of moults that, from hatching to the final product, are very similar to the adult insect, though only adults have fully developed wings.

Grasshoppers have very stretchy abdomens which allow them to place eggs deep in the ground.

You can witness the life cycle first-hand easily by rearing some in captivity. Collect some nymphs in spring or summer and keep them in a well-ventilated vivarium. A plastic aquarium with fine netting on the top is fine. They do best if you provide your own 'sun' – don't leave them on a sunny window ledge or you'll fry the inhabitants. Instead use a 60-watt bulb on a timer, mounted on the inside of the tank, or a desk lamp held over the set-up.

When feeding, tie a small bundle of mixed grass together on a piece of string. This can be lowered into the tank without the risk of the nymphs jumping out, and makes changing the old food easy too. Grasshoppers can be fed a varied grass diet, either freshly cut or planted in soil. Crickets, though, need plenty of variety – bread soaked in honey, grasses, flowers (good source of protein), fish food flakes, fruit and live insects such as young blowflies (easily obtainable from fishing shops).

Provide some nice twiggy branches; this gives the insects something to cling on to with plenty of height for when they moult. This is a struggle that must be seen to be believed – such a complex and leggy beast, hanging upside down, splitting at the neck, and pulling its delicate pale body, including antennae, out of the old skin. Depending on the species they will shed their skin between four and 10 times before reaching adult size.

This is a cricket using the privacy of darkness to go through the vulnerable process of moulting its skin.

Adult females will readily lay eggs in damp soil or sand, and it's fascinating to see them do this. Some take several years to hatch, though, which makes the keeping of them all the way through their life cycle a bit of a challenge for all but the most dedicated grasshopper and cricket fan! Having said that, you can certainly learn a lot about the ways of adults by keeping a few together. You could well get to see them singing up close, while watching a bush cricket ambush its prey is every bit as exciting as a lion running down a zebra on the Serengeti. This is the kind of action you rarely get to see in the wild.

The first time I kept a pair of Great Green Bush Crickets and saw them mating, the male produced a white, creamy mass containing the sperm. This was attached to the female in a rather humiliating process for the male, as he was dragged around and hung upside down for the best part of 20 minutes. Finally when he got off, she proceeded to eat some of the mass, the nutritious meal would help her reach peak condition to produce healthy eggs.

Beetles – so many species, so little time

When biologist J. H. Haldane was asked about what he could say about the creator by studying living things, he replied that the creator had 'an inordinate fondness for beetles'. This quote has become the credo of coleopterists everywhere, and I'm not ashamed to repeat it here.

If nature was an inventor, beetles would be her most successful production, a design classic. With more than 350,000 species named so far, they make up more than a quarter of all animals on the planet. They've been around for a while too, their body-plan having had 230 million years' worth of testing by nature's R&D department.

From the tiniest, less than a quarter of a millimetre in length, to the gargantuan 20cm (8in) beasts that are the Goliath beetles, they all have something in common. The probable secret to their success are the wing cases or elytra that most beetles have on their back. These highly modified forewings act as protective covers for the delicate, folded flying wings.

A lovely example of a well-named family of beetles, the jewel beetles.

Most beetles can fly, so can spread out and find food resources over great distances like other winged insects, such as butterflies and dragonflies. They may not have much finesse in the air but what they lack in grace they make up for in practicality. They can fold their wings up and put them away, and go to places that other, more delicate flying creatures cannot. Imagine the mess a butterfly would get into if it tried to dig under a cow pat or just crawl down a hole – it simply couldn't do it.

Next time you find a ladybird or a chafer beetle, persuade it to climb onto your hand. It should crawl to the highest place and sit poised on your fingertip, 'thinking' about flight. You'll see those elytra spring open, and after a pregnant pause the transparent hindwings are unfolded on their delicate micro-hinges, briefly stretched, and the beetle goes humming off. Beetles also have simple mouthparts, which means these and the rest of their body can be customised, moulded, extruded and modified. Shake a bush, dip into a pond, kick open a cowpat or dig in the soil and you'll probably find a suitably adapted beetle.

With the exception of the long and slinky rove beetle, whether a minuscule Flea Beetle or a great big shiny behemoth of a Goliath Beetle, the overall anatomy is pretty much the same, a tried and tested design, solid and robust. The basic body-plan has be shaped, moulded and stretched to many different ways of life.

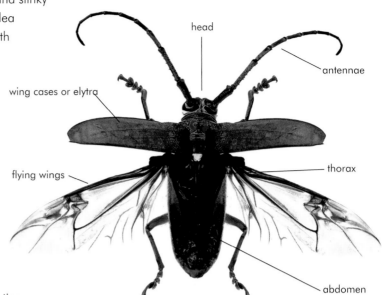

head

antennae

wing cases or elytra

thorax

flying wings

abdomen

Most can fly although there are some which have lost this power and their wing cases are fused. When functional wings are present it is always the rear pair which are used for generating propulsion through the air – the first pair, the wing cases or elytra, are held high and out of the way; their job is to protect the wings when not in use. Just like a wind surfer's sail the wings have a complex series of hinges and folds which allow them to be packed away when not in use, something that you might just get a chance to witness if you watch a beetle the second it lands.

Being insects their body is divided into the three parts we come to expect: head, thorax and abdomen.

Beetles feed on many different things and their mouthparts reflect this; most have chewing and biting mouthparts that move from side to side but they vary in size, strength and length. Those that chew through wood and plant material such as the long-horn beetles are equipped with mandibles like tin-snips, while others chew and mash softer foods like pollen and rotten fruit. Some such as Stag Beetles have turned their mandibles into combative tools, for battling with rival males, and so they only have a 'tongue-like' galea which is used to sop up sweet liquids like sap and fermenting fruit juices.

wing cases or elytra

The often overlooked life-cycle of the ladybird. All beetles have such a life-cycle – but not all are quite as bold and easy to find, if you know what you're looking for.

1. Seven-spot Ladybirds mating

2. A batch of ladybird eggs

3. Seven-spot Ladybird larva – always looking for dinner

4. Ladybird pupa

5. Ladybird freshly emerged from pupa (right); the beetle is very soft and pale and it needs to rest and harden up

6. The finished beetle all hardened up and in full colour

Beetle life cycle

Like butterflies, beetles undergo complete metamorphosis in their life cycle with egg, larva, pupa and adult stages. You can easily investigate the beetle life cycle in captivity by rearing mealworms or fruit beetles. You can also rear several of the common ladybird species that live in gardens.

A few adult ladybirds can be kept in plastic Petri dishes or any clear plastic box, lined with a piece of kitchen towel to absorb any excess moisture. Keep them well fed on live aphids, and they will in time lay clutches of yellow skittle-shaped eggs, from which tiny, voracious larvae hatch. Keep no more than five of these and release the rest on aphid-ridden plants outside. Ladybird larvae are top-notch predators and while they prefer aphids, if they run short of these or are overcrowded they will turn cannibalistic. Clean them out every day and place fresh aphids in their pots on leaves. Gardeners in the family will probably be very interested to watch as the baby beetles make mincemeat of their arch-enemy.

There are four skin changes as they grow. Just before the larva moults into the pupa, it stops moving, attaches the tip of its abdomen to a surface and hunches up. This is the pre-pupa – 24 hours later it will moult for the last time, becoming a strangely active pupa, which will 'stand up and down' several times if it is touched. It's the same colour as the larva, but still doesn't look like a ladybird.

Two weeks later a pale beetle will crawl out, although it still doesn't look like a ladybird. It will rest for a while, allowing its wing cases to harden and develop colour. Now it's clearly the insect we know and love.

A micro-monster – watching a ladybird larva chew through a colony of aphids reminds us they're not as nice as their public persona might suggest, you almost feel sorry for the aphids.

Many ladybirds form aggregations in which to pass the winter. This spectacular cluster is formed by Seven-spot Ladybirds.

~ CATCHING BEETLES OUT: PITFALL TRAPS ~

This is a method for catching some of the most spectacular insects, the ground beetles (carabids). You will also undoubtedly catch many other kinds of insects too, and if you use meat as bait you may get burying beetles and the elongated staphylinids, such as the spectacular Devil's Coachhorse (*Ocypus olens*).

Any deep canister with smooth, slippery sides will do. The most basic design is a tin or jar sunk into the ground, with a few dry leaves in the bottom and a stone or tile placed over the top, balanced on a few small sticks or stones. You can place these anywhere in the garden – beetles on their hunting forays may then fall in. Or take the design one step further and create an even more effective trap.

You will need:

- three sticks
- string
- 1 large and 1 smaller plastic drink bottle
- skewer
- scissors
- leaves and twigs
- trowel
- very ripe bananas
- sugar

1. Take the larger drinks bottle and cut its neck off with the scissors. This is going to be the actual pit. To ensure it does not fill up with rainwater and drown the beetles, use the skewer to make plenty of drainage holes in the bottom.

2. Dig a hole deep enough to sink the 'pit' into, and place it in the ground so that the top is level with that of the soil around it.

3. Take the smaller bottle and carefully make a good number of holes with the skewer in the sides and top of the top two-thirds (not the bottom part).

4. Now prepare the bait. Fermenting bananas are one of the best – many professional beetle hunters will back me up on this. So preparation of this a day or so before you go trapping is a good idea. Squeeze the bananas piece by piece into the top of the second bottle. Add a bit of sugar, perhaps some yeast, and leave in a warm place for a few hours.

5. Back at the trap site, make a wigwam out of sticks tied together at the top; place this over the trap. It must be high enough to allow you to suspend the small bottle with the bait in it over the top of the trap, without touching the ground. Put some dead leaves, twigs and bark in the bottom of the trap, but not so much that insects could climb them like a ladder and get out; you want them to fall in, then hide and be safe until you find them.

6. Check the traps at least daily and always first thing in the morning; some of these beetles if left in the traps for too long may start to snack on their smaller, trap-mates.

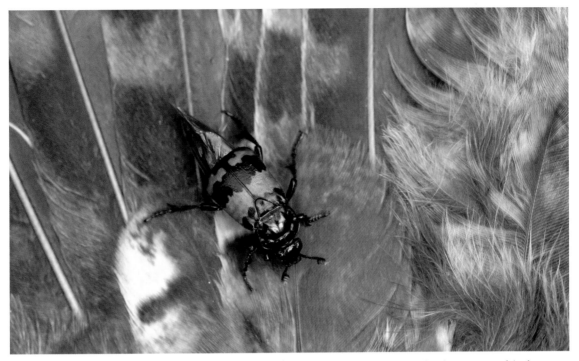

Sexton beetles are some of the first on the scene when an animal dies – they really are nature's undertakers.

Nature's undertakers

If your cat brings in a dead small mammal or you stumble across one in the field, you can use it to lure in those great recyclers, the sexton or burying beetles. Peg down the carcass under some chicken wire, or better still place it under a piece of corrugated tin or a flowerpot. Soon it will begin to decompose a bit, not necessarily enough for you to smell – but if you were a burying beetle sensitive to the smallest quantities of the odour of putrefaction it would get your antennae twitching.

They get on the job by flying – their incredible sense of smell draws them in over a great distance. Once on the carcass, they fold their flying wings under the protective wing cases and get to work burying the body.

Several different species could turn up at the bait, from big black burying beetles to a few species with bright red bands on the elytra. The first male to arrive has a job on his hands – he must defend the body against other male beetles while waiting for a female to share his spoils. It's first come first served, and the congregated beetles will fight until the strongest pair is left. They will then bury the corpse. Under the ground, they will shave and mould it into a perfect serving. They will also mate and lay eggs nearby, at the same time removing and eating any fly eggs that may have been carried down with the corpse.

Leave a light on

Some beetles have developed the 'technology' to light up parts of their abdomens, to attract a mate in that huge sea of blackness that is the night. The light show put on by fireflies and glow-worms is seen at its best in the tropics. (Normally these names actually refer to beetles despite the names. There are some midge larvae that glow, also called 'glow-worms'.) Many pleasant sundowners have gone on a little longer than intended because when the sun sets the beetles rise. The glow is caused by an enzyme called luciferase acting on another chemical, luciferin.

The luminosity is mainly to attract mates. In some species, such as the European Glow-worm (*Lampyris noctiluca*), the flightless females glow to attract the winged males, which are searching for just this beacon of love. Although sadly a rarer experience than it once was, it is still possible to go out on a warm summer's evening in parts of Europe and find a grassy bank with a glowing constellation of these wonderful insects. In other species both sexes glow, engaged in a kind of sexy semaphore.

If you investigate the vegetation nearby using a hand torch, you may find a male that has flown down to investigate the female's lantern. You can, if you've got an LED torch of the same or similar colour, attract the males down to your super pseudo-female.

This female Glow-worm keeps her lamp light burning in order to attract the winged males.

In other parts of the world the most visible fireflies are males of numerous species, which either perch or fly around, flashing their signals to others of the same species. The colour intensity and frequency of flashes is species-specific and when a female, usually resting on some vegetation nearby, gets the right 'Morse code' she responds with her own flashing signal, giving the green light to the male. With a little practice, by copying the female's response signal you can lure the males to you instead! You must keep your interference like this to a minimum, just enough to top your soul up with awe and wonder, then it's lights off to let them get on with the night's serious business.

Ants, bees and friends

From the windowbox outside an office in deepest Manhattan to the rainforests of Brazil, the Hymenoptera are doing what they do best – keeping busy. We get hung up on the ones that make honey, pollinate our crops and sting us, and tend to overlook the other vital roles these insects play in the running of the world. Members of this order crop up in every habitat on earth that isn't frozen solid. In many places they are the most numerous living things other than the plants themselves; the ants alone in a given area of rainforest would outweigh all the other animals together. There are more than 115,000 species in all, ranging from large and showy wasps and bees to the ant workforce slaving away underground, in dead wood and between cracks in the paving slabs.

This White-tailed Bumblebee is busy collecting pollen and nectar and in doing so ensuring this fruit tree is pollinated, one of the many 'services' these insects perform.

Hymenoptera is roughly split into two groups – the ants, bees and wasps (suborder Apocrita) and sawflies and woodwasps (suborder Symphyta). The name Hymenoptera means 'membrane wings'.

We exploit some of their many talents commercially, not only in the production of honey by Honey Bees (*Apis mellifera*) but also pollination services in orchards and greenhouses by bumblebees and other species of bee (bees from a single nest will make between two and three million visits to flowers a day). Flowers and social pollinating insects co-evolved; without the insects we would probably not have flowers as we know them, and therefore no fruit. Remove the social insects and life as we know it would rapidly grind to a halt. We also harness the grisly efficiency of parasitic wasps as biological control agents against many glasshouse pest species.

Parasitic wasps while not always as obvious as their social cousins perform important jobs controlling pest species and keeping things in balance.

Getting to the point

We are only too aware that these animals have a sting, so let's get to know this part of the anatomy a bit better – armed with knowledge maybe we won't fear it as much. The sting is actually a modified egg-laying tube, or ovipositor, evolved to allow a female to pierce an animal or object to lay eggs in it. Some of the parasitic wasps still use theirs in this way. Most stinging Hymenoptera only use the sting for defence, though a few solitary wasps also use it to paralyse their prey. Although only females carry stings, almost all ants, bees and social wasps are female – because of the way their lives work, males exist only at certain limited times of the year.

Many people swear they have been attacked in an act of calculated malice by a wasp or bee. They are wrong – let me explain why. The essence of a sting is manufactured in the venom sac and is costly to produce (the equivalent in energy of you or me running several times around a football pitch for no reason). By deploying it, those species with a barbed sting (Honey Bees being the famous example) pay the ultimate price of death; the sting gets snagged in the flesh and is torn out along with the venom sac, fatally wounding the bee.

The sting of a wasp is just a modified egg-laying device!

If you are stung, do not pinch the sting out with your fingers, or you'll squash the venom sac and squeeze more venom into the wound. Instead, scrape the sting away with your fingernail. It's most likely to happen if you disturb social bees or wasps that are defending a large colony or stored honey. I once stood up under some low scrub on an Amazonian river bank and ended up wearing the nest of a colony of understandably angry wasps on my head like a Chinese hat! A few of the parasitic wasps will deliver a painful sting as well, as I've found to my cost when trying to get them out of a moth trap or window.

The biggest danger from any insect sting or bite is individual sensitivity to the venom. In extreme cases the reaction may bring about anaphylactic shock. If you are one of these unfortunate people, it's wise to carry an epipen. But none of these insects will sting unless provoked – the only exception I know of (due to painful experience) is if you stand directly in the flight-path approaching the nest of social wasps or bees. Just don't do it.

One of the most efficient ways of utilising the space available, the honey comb of hexagonal cells was created by the Honey Bee.

Honey Bee queen cells and pupae.

The social whirl

Bees and wasps have a life cycle similar to that of butterflies and moths: they have an egg, a soft fleshy larva, a pupa and a winged adult stage. The pupa is rather untidy compared to that of a butterfly, with lots of bits like legs, antennae and wings sticking out. We usually see only the adults, as the other stages are usually locked up inside the nest chambers, cells or combs.

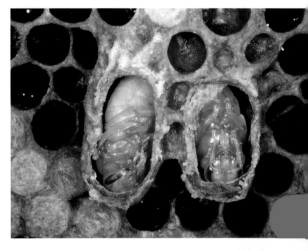

There are two common sorts of social bees that buzz round our gardens – the bumblebees and the Honey Bee. Bumblebees are the round cuddly ones, brightly striped in yellow and black, and all the more noticeable because they appear in early spring before many other flying insects. The Honey Bee is slimmer with less bold stripes. Throughout the world they gather nectar from which they manufacture honey, pollinating our flowering plants in the process and therefore playing a key role in life on earth as we know it.

A Honey Bee with visible pollen sacs in flight.

Unlike bee larvae, which feed on pollen and nectar, wasp larvae are carnivores, and are fed on all the things that gardeners hate – greenfly, caterpillars and other insects that munch their way through vegetable patches and herbaceous borders. Adult wasps, like most of the Hymenoptera, have a sweet tooth, which has the downside that they are always keen to have a lick of your ice-cream when you are out in the park on a sunny day. On the other hand, wasps are just as good at pollinating as bees are. Don't write the wasps off as nature's bad guys – they perform vital services too.

Communal living

Ants and social bees and wasps live in colonies, composed mainly of female workers, a sisterhood, co-operating in all the tasks necessary to further the life of the colony as a whole. Each nest is presided over by one or more queens (depending on species), which may live for one year or several.

Bumblebees and Honey Bees have somewhat different life styles, but both their communities revolve around the queen. She is a highly efficient egg-laying machine and nest manager. Honey Bee colonies live for several years and survive the winter and any other harsh times by living off reserves of honey, while bumblebees have evolved to complete a colony cycle inside a short period of time, with only young mated queens hibernating through winter.

In a Honey Bee hive there are anything from 20,000 to 80,000 bees, all the offspring of one queen. Some of the members of the colony are males, known as drones, whose function is to mate with any virgin queens they find while out flying in the sunshine; after this they die. But the vast majority are workers, infertile females who look after the hive and the larvae, producing the royal jelly on which the young are fed for the first three days and foraging for the pollen and nectar most larvae will eat after that. Those larvae designated as future queens will continue to be fed on the protein-rich jelly. When a queen emerges from her pupa, she leaves the hive, mates with males from other colonies, then takes a task force of workers which form a swarm and leave to set up a new colony. Colonies will thrive as long as there is a good supply of honey and a healthy queen.

Bumblebee colony life is slightly less organised, with smaller, less orderly nests and a different approach to life, but they are beautifully successful in their own way.

The social wasps and ants have different strategies again. The paper wasps – the classic yellow and black stripy insects we're most familiar with – have (for the most part) an annual nest cycle. The colony is founded by a single mated female, it builds through the season until the males and next generation of virgin queens are produced, then they all leave, mate and the cycle starts again. Hibernating queens may be found under loose bark, and in lofts, garden sheds and greenhouses over the winter.

Ants start off with a single mated founding queen (who will hibernate in temperate climates) and the colony builds around her. Many species have several long-lived

These wood ants form mounds as a way of controlling the temperature within the nest. There will always be a hot side and a cool side; also the mass of ants themselves generate heat and in the spring their bodies form a dark mass to help them absorb the first of the spring sunshine's warmth.

The highly organised colony order is all down to the queen who controls everything remotely using hormones and other cues.

Not as organised as the nests of Honey Bees, the smaller colonies of the bumblebees are a messy cluster of brood cells and storage cells. It works for them.

queens and these along with the colony can survive the harsh months by going into periods of inactivity deep underground. Some species even store food resources, such as seeds, deep within the nest. Honeypot ants store food in their own workers, and many common garden species such as the Pavement Ant (*Tetramorium caespitum*) can store food and water in the 'crops' of the individual workers, creating a kind of communal stomach.

The singletons

Most bees and wasps are solitary species. They create single lined cells, or burrows, which are provisioned with food, either pollen or paralysed prey. An egg is laid in each cell, the larva feeds on the stored food and usually hibernates as a pupa, before emerging the following year to mate and start the process again.

Some species makes their presence known by their method of nest cell construction. Masonry bees will collect mud of a specific consistency to partition and plug their nest cells, while leafcutter bees fill their nest burrows with a series of cells made from discs of plant leaves, neatly clipped from various plants, leaving the leaves of many a rose bush with neat circular bits missing. Others go to great lengths to find empty snail shells which they stuff with a pulpy mastic of leaves and mud, adding their eggs and pollen before covering up the whole lot with a thatch of twigs!

Although each nest is individual, these insects often nest close together in colonies. A sandy bank or hot sunny wall or fence post can be abuzz with the comings and goings of dozens of individuals of different species.

Some fire ant colonies contain many queens and number over half a million insects! Their power really is in their numbers for these small ant species.

The majority of ant, bee and wasp species are like this Ashy Mining Bee – they live alone.

Attracting bees

It's well known that bees are declining. To help bumblebees – and benefit from their pollination in your garden – you can encourage them to take up residence by recreating the conditions found in rodent burrows, a favourite place for many species to nest (see below). Or you can build something that looks like a garden feature but is really a useful bee-nesting site – bare soil is a rare habitat in nature. Heap soil up into a bank, somewhere where it will get sun for most of the day. The soil needs to be what gardeners call free-draining, so if yours is a bit muddy, mix it with sand or gravel. Weed at least some of the bank so that bees don't have to fight their way through the vegetation, and you will soon see them starting to excavate burrows. A pile of logs stacked on top of the bank, with holes of different sizes drilled into the ends and sides, will attract those species that like to live in wood.

It also helps to provide pollen and nectar sources for the adults (which will in turn benefit many other species). Choose plant species that are 'wild-type', rather than primped and overbred varieties that are about aesthetics rather than nectar and pollen production. Herbs such as sage, thyme, rosemary and lavender are good to start with, and there are loads of website resources that will help point you in the right direction. What will work best for you depends on the soil type, climate and the habitats in which you and the bees live.

The easiest and simplest way to attract pollinating bees to your garden is to make a nestbox for tube- and cavity-nesting species such as the harmless Red Mason Bee – which bee for bee is a better pollinator than other better known bees such as Honey Bees and wasps!

~ MAKING A BEE BOX ~

The addition of a bee box to a nice sunny wall will bring the buzz of life to even the smallest concrete garden – all you need are a south-facing, sunny wall and some nearby flowers. These boxes are most attractive to little mason bees of various species. They're very reluctant to sting (some can't, while others don't have the hardware to get through) and they are excellent pollination value – for each female you would need 120 Honey Bees to do the same job!

These bees can be really fussy, so increase your chances of success by making a number of boxes and set them up in different parts of the garden. Having said that, I have had the poshest-looking professional bee box in my garden for some years – short of providing an en-suite bathroom, I don't know what else I could do to attract them – and only spiders, earwigs and a few beetles ever use it. I also have a version I made myself from an old catering-sized coffee tin and a bunch of old buddleia and elder stems, and it's a positive 'hive' of activity.

You will need

- hollow twigs and plant stems of different diameters; for example hogweed, elder, grasses and reeds. You can also add bamboo canes and paper drinking straws
- flowerpot, coffee jar, wooden box or similar container
- scissors
- string, wire or twine

1. Fill the flowerpot or whatever with a selection of stems, trimming them to the height of the container.

2. Attach the pot to the stem of a shrub, tree trunk, wall or fence in a sunny south-facing spot. Make sure the entrance tilts down slightly, so that it doesn't get waterlogged when it rains. Nests in position in the early spring and summer will do best.

3. If your housing is approved, you'll soon see lots of solitary bees coming and going with building materials; many will be various masonry bees (*Osmia* sp.) but all manner of other solitary bees and wasps (including the parasitic cuckoo wasps) and other insects may turn up as well.

Thanks to the current trends in wildlife gardening and the recognised pollination services these wonderfully fascinating, benign insects bring, you can buy bee boxes in several designs. One of the best purchases I've made for my garden recently is a woodcrete bee box with clear tubes and a removable front, so you can watch activity inside.

BUILDING A BUMBLE ABODE

Those first bumblebees that you see cruising around in early spring are the new queens, who will soon be looking for a suitable hole or crevice in which to nest. It's your aim to impress them. The more nestboxes you build and install the more likely you are to succeed. Queen bees are fussy creatures. The perfect burrow must be cosy, well lined and ventilated. There are many different designs that you can use to try and attract bumblebees to your garden – below is one of the simplest, but feel free to experiment. A cautionary note: don't use cotton wool for your stuffing material – bees catch their feet in it.

For each nest you will need:

- small flowerpot with a single drainage hole in the bottom
- some fine chicken wire
- strong multipurpose glue
- small section of fine mesh, gauze or muslin
- length of pipe (approx 18mm (¾mm) in diameter)
- 4 corks or similar-sized stones
- piece of wood or an old roof tile or slate
- sawdust, hay, dry moss, kapok stuffing and old mouse bedding
- trowel

Step 1

1. Glue a small square of the fine gauze or muslin over the hole in the bottom of the flowerpot – this makes it less likely ants will invade the nest.

2. Dig a hole in a sunny spot in your herbaceous border, at the edge of the vegetable plot or even on the edge of the lawn where it won't be disturbed by a lawn mower. The hole should be broad enough to accommodate your flowerpot, and a little deeper, with a small trench to bury the entrance pipe.

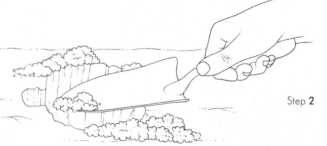

Step 2

3. Place a pad of bent-up chicken wire in the bottom of the hole to help air circulate around the nest chamber. On top of this put a mixture of the stuffing materials, about enough to fill the flowerpot. The addition of old mouse bedding is a little secret ingredient; the questing queens can smell old mouse nests. Pet shops are a reliable source if you don't know anyone with a pet mouse.

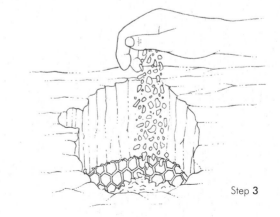

Step 3

4. Now rig the pipe so it comes up through the chicken wire, and then runs up to the surface.

5. Place the pot upside down over the contents of the hole – the bottom should be more or less flush with the surface of the soil. Now cover the drainage hole with the tile, slate or piece of wood, balancing it at the four corners on the corks or stones.

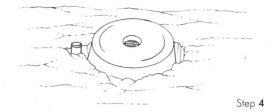

Step **4**

6. At the end of the year, when all activity has ceased, gently lift the flowerpot and you should see the remains of the wax cells made by the bees. You may even find a new queen or two, sitting out the winter in your cosy nest. If you do, keep disturbance to a minimum, put everything back in position and you may well have a repeat next year. Watch out for bees coming and going through the summer.

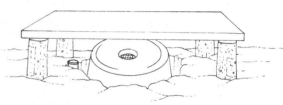

Step 5

I've had success using cleaned-out old paint tins, covered in cement and hidden around the garden; basically anything that might produce a cosy, dry and dark habitat similar to a small mammal burrow can be utilised. Alternatively, you can buy purpose-made bumblebee boxes from garden wildlife suppliers.

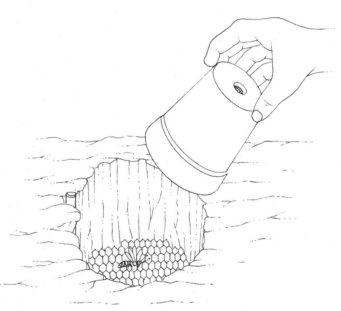

Step 6

~ BEE-TRAINING ~

This lovely experiment is perfect for a lazy summer's afternoon in the sunshine. The bees get a nice sweet reward, you get close-up views of these insects in action plus you get to learn a little about how Honey bees and other insects perceive the world around them. First you need to train your bees.

You will need:
- pieces of different-coloured card
- 4–5 identical bottle caps, or better still watch glasses
- sugar (white, refined) solution
- paintbrush and some non-toxic quick-drying paint

1. Cut identically shaped flower outlines from the different-coloured cards. Mix up some sugar solution with a little water.

2. Line the 'flowers' up in the sun on the grass or a table, making sure they don't catch the breeze and blow away. Place a bottle cap or watch glass in the centre of each one.

3. Add your sweet solution to just one of the caps.

4. Now you have a choice. You can simply wait for a curious bee to land on your flower and find the reward for itself, or you can soak an artist's paintbrush in some sugar solution and go and see if you can gently persuade a Honey Bee to step off a flower and feed from the bristles. You can then place the bee on your 'full' flower and get it to start feeding there. Now is the time to dab a tiny bit of coloured paint onto the thorax of your bee – you'll then recognise her if you see her again.

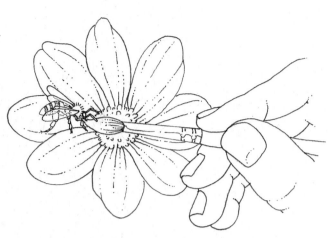

5. Watch and wait. Your bee will imbibe the sweet solution and you can study her and watch her tongue in action. When her stomach is full, she will fly off – not in a straight line, but she'll circle around to get her bearings before heading off to the hive.

6. Now wait – depending on how far away her hive is, she will return; you'll recognise her by her paint dab, and she'll almost certainly be joined by a few of her nestmates. She has passed on the directions to your flower back at the hive by engaging in a beautiful little bit of behaviour called a 'waggle dance'. She has also shared a little of the food with her sisters and they're back in force – their numbers will keep building over time. This is how Honey Bees get so good at exploiting a wild food source. As soon as a plant, bush or tree comes into bloom the bees plunder its nectaries with great efficiency.

7. Now you can play around with all sorts of different questions and experiment a bit. If you add honey solution to all the 'flowers', what happens? What if you move the original 'flower' away from its first position – do the bees still find it? You could experiment with different colours – are some found quicker than others? How long can a bee remember the colours that rewarded it, if you remove all the 'flowers' and then replace them again after a given period of time? Do the bees prefer different strengths of sugar solution? Oh, and in case you're worried, feeding bees are very passive and as long as you don't squash or squeeze any you'll not get stung!

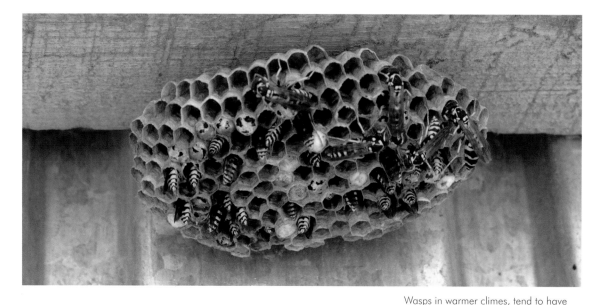

Wasps in warmer climes, tend to have exposed brood combs.

Wasp nests

Of all the social insects, the social wasps get the worst deal. They don't sweeten things for us by storing honey, instead we worry about the sharp end with which they vigorously defend their colonies. Often I'm asked the question: 'What's the point of wasps?'. Well, they inspire and intrigue with their amazing behaviour and ingenious lifestyles, they are excellent pollinators (as good as bees) and they are also effective and systematic pest controllers (better than any ladybird).

Keeping and attracting them should be as commonplace as with bees, in my view. Providing wild banks where rodents can live works for wasps as well as for bumblebees with the provision of flower-rich borders. Cavities such as birdboxes work for tree-living species; an early queen wasp has hijacked many a birdbox. If this happens, don't panic, try to enjoy the pleasure it brings.

You may well occasionally witness wasps' outdoors behaviour; most commonly the collection of wood fibre. Our attention is initially grabbed by the sound of scraping and scratching as a worker wasp goes about the collection of nest material – dead wood or plant fibre. This is mixed with saliva to make a wet pulp, which is then laid down to form the 'carton' with which the nest is fabricated. If you are lucky enough to come across an old

This queen Hornet has just started building her nest and colony – the eggs will hatch and be her first workforce that will grow as the seasons progress. In the meantime she'll incubate these eggs by vibrating her flight muscles to keep her daughters warm.

or abandoned nest, look carefully at the exterior construction and you'll see it's made up of slightly different-coloured bands; each separate colour represents a different source of wood fibre. Also note the brood cells, sharing their geometry with the hexagonal wax cells of Honey Bees (this shape is not only the most efficient use of space, but is also the natural default pattern when soft/wet pulp or wax makes contact with neighbouring flexible cells – think of the shapes of bubbles in your bubble bath when they meet). The main difference (other than the construction materials) is the plane in which they are built – wasp broods are almost always horizontal while bees' are vertical.

In very hot weather you'll often see wasps (and bees for that matter) gathering around sources of moisture, be this ponds, puddles or bird baths. Here they are collecting a gut full of water to spit out on the inside of the nest, where it will evaporate and cool the nest interior.

While this wasp looks like she's having herself a drink, she may well be using her stomach for the community and taking water back to the nest – to share with others or to increase the humidity or cool the interior of the nest.

Ants: a great success story

There are a lot of them – some 14,000 different species, with more being discovered all the time. They have been around for an impressive 80 million years and they are probably the most successful living creatures in the world today. They are basically wingless wasps, and unless you have a trained eye and a good microscope many of them look very much the same.

So what makes the ants so successful? Their small size is one factor, but the other is that they are highly social and live in huge, terrifyingly efficient colonies. These colonies are ruled over by one or more queens, depending on the species. In addition to laying all the eggs, the queen exudes chemicals called pheromones that are passed from ant to ant in the food they share and in their meeting and greeting process, which includes lots of licking and stroking with their antennae. These pheromones control the colony, sending out the queen's instructions about defence, foraging parties and other vital activities. The inhabitants of a colony are all the queen's daughters, who do not

Using the power of numbers these ants form a bridge from their own bodies... just one example of the sorts of amazing things you'll witness if you study these incredibly important little insects.

breed themselves but devote their energies to servicing the queen and the community. They feed the larvae on nectar and insect grubs, moving them from room to room in the colony as they grow; when the larvae pupate the workers move them to the top of the nest where the warmth of the sun will help them mature quickly; and they go out in huge parties to search for food.

The Giant Leafcutter Ant queen dwarves her daughters; she's one big egg-laying factory laying up to 30,000 eggs a day.

When the time is right, the reigning queen releases special hormones that allow her eggs to produce winged virgin queens and males instead of the constant stream of female workers. One hot sunny afternoon in the middle of summer, something – it may be temperature, day length or humidity, no-one really knows – triggers a colony to send its potential new queens into a glorious nuptial flight. Mating takes place in the air. The males, having served their purpose, then die and a newly fertilised queen either returns to her former colony or, more usually, sets out to find a suitable site in which to found her own. She immediately starts laying eggs, living off her fat reserves until the first clutch hatches into workers ready to start feeding her again. She uses her jaws to pull her wings off – they too have served their purpose; she will literally spend the rest of her life laying eggs.

All ants produce winged royals. These virgin queen and male fire ants are about to swarm out of the nest. They will mate, the males will die and the females will go on to found new nests.

~ ANT TOWN ~

Ants do most of their social living underground, so building an ant city is the best way to observe them in action. To collect your ants, look for nests under stones and wood. Early spring is the best time, as the ants cluster together and you can easily gather a scoopful, which makes it more likely that you will harvest a queen and a good selection of workers, eggs and pupae.

You will need:

- piece of plywood 30cm (12in) square for the base
- 6 strips of wood about 2cm square in cross section for the frame
  – 1 × 28cm (11in) long, 2 × 24cm (9in) long and 3 × 10cm (4in) long
- Plasticine
- piece of clear rubber tubing at least 30cm long
- sheet of hard, clear plastic 28cm square
- plaster of Paris
- well-drained soil from the garden
- jam jar with a lid
- enough black cloth or paper to cover the formicarium

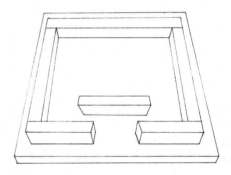

Step 1

1. Screw the wood strips to the base to make a frame. The longest piece is the back, the two medium-sized pieces are the sides and two of the short pieces make the front, leaving a gap. Put the final short piece of wood inside the frame, parallel to the front, but don't screw it down.

2. Use the Plasticine to form tunnels and chambers the same depth as the frame. Lay the sheet of plastic over the top to test this – if the Plasticine squishes against the plastic, you have built your ant city too high. Make sure the Plasticine snakes around the loose piece of wood. Fill the entrance with more Plasticine and push one end of the rubber tubing through it, butting it up to the loose wood.

3. Mix a runny solution of plaster of Paris (see p.85) and pour it into the gaps not filled by the Plasticine. Leave to set for a day or so, then pull out the Plasticine and the loose piece of wood to leave a series of tunnels in the plaster of Paris. Fill the cavities left by the Plasticine and wood with soil.

4. Make a hole in the lid of the jam jar and attach the end of the rubber tubing to it, making sure it is flush with the inside of the lid and sealing any gaps with Plasticine.

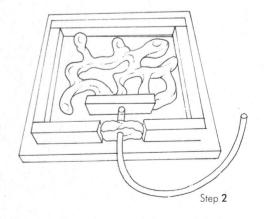

Step 2

continued overleaf

5. Before introducing the ants, place your formicarium in a warm location out of direct sunlight and cover it with the black card or cloth. Now put the ants into the jam jar (which should remain exposed to the light), add a little cotton wool and soak it in weak sugar solution. Put the lid on firmly and secure it with tape to make it completely ant-proof.

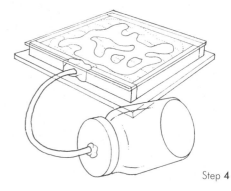

Step **3**

6. After a brief period of chaos, the ants will get their act together and move into the formicarium. Leave them for a few days to settle down, then start putting different types of food into the jar and watch as the ants march back and forth to collect it. They need sugar, water and protein, which you can provide with honey, damp cotton wool, insects (alive or dead – a neglected windowsill is a good source of these) and seeds. Replace the jam jar with another one whenever you need to clean it out and replenish supplies.

If the ants you collected included a healthy queen, the numbers in your colony will soon start to increase as she lays more eggs and the workers rear more recruits.

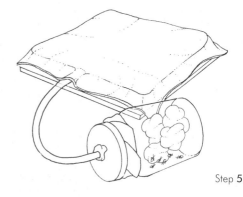

Step **4**

Step **5**

Pavement Ant adults with larvae in their nest.

Arachnids – eight lovely legs!

The arachnids are forever being lumped together with insects, not helped by the fact that many insect field guides have a section on spiders in the back. All you need to remember is anything with more than six legs is not an insect, and that includes all of the animals in this group, the arachnids.

But the class Arachnida holds much more than spiders alone. There are 11 subclasses, amounting to an incredibly diverse 80,000 species in all, from the huge Goliath Bird-eating Spider (*Theraphosa blondi*) to the smallest mites and ticks. Among their number are some of the most sinister-looking and efficient predators.

They all have a body divided into two main parts: a cephalothorax and an abdomen. Attached to the former are four pairs of walking legs, a pair of pedipalps (leg-like in spiders, and pincers/claws in scorpions) and chelicerae (the piercing fangs of spiders and micro-pincers of scorpions). The 11 subclasses are:

1. Spiders
2. Mites and ticks
3. Scorpions
4. Pseudoscorpions
5. Sun spiders
6. Whip scorpions
7. Microwhip scorpions
8. Short-tailed whip scorpions
9. Tail-less whip scorpions
10. Hooded tick mites
11. Harvestmen

Two spectacular arachnids – A Mexican Red Knee Tarantula (*Brachypelma smithii*) and a Madagascan Scorpion – although most arachnids are much smaller.

Spiders

Arachnophobia (or 'Miss Muffet syndrome') seems an almost fashionable fear nowadays. Very few people are actually truly phobic, but all you have to do is mention the word 'spider', let alone reveal a living one, and you are pretty much guaranteed to send someone flying out of the room in hysterics!

Of 35,000 species in the world, only 20–30 species are dangerously toxic to humans and even these very rarely kill. For every reason we give for placing them under the foot, there are many good reasons to like them too. Their variety and diversity alone is stunning – they come in many different shapes, sizes and designs for different ways of life. Some hunt on foot, some spin elaborate traps. There are underwater spiders, spiders that give gifts, spiders that nurse, surfing spiders, parachuting spiders, jumping spiders, spiders that steal from other spiders and even pirate spiders that murder their fellows. They are a truly versatile bunch, but all have a number of things in common.

Above from left to right: Jumping Spider, Crab Spider and Orb Spider.

Who said spiders were ugly? Few invertebrates show such a diverse array of lifestyles and hunting techniques and quite a few are rather pretty as well.

Know your way around a spider

The best way to learn about spider anatomy is to catch one and investigate it, using a hand lens and a bug restrainer as most species are very active and will not give you much observation time before putting those eight legs to good use. If you can't find an actual spider, find an empty moulted skin instead. This is a great way of learning about the anatomy of these animals, without the living spider actually being inside. You can find one of these in pretty much any neglected corner of the house (try the cupboard under the stairs or a dark corner of the garden shed).

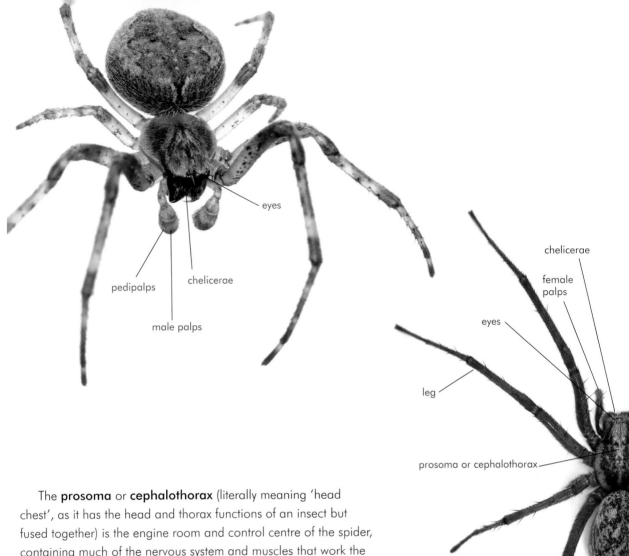

eyes

pedipalps

chelicerae

male palps

chelicerae

female palps

eyes

leg

prosoma or cephalothorax

abdomen or opisthoma

spinnerets

The **prosoma** or **cephalothorax** (literally meaning 'head chest', as it has the head and thorax functions of an insect but fused together) is the engine room and control centre of the spider, containing much of the nervous system and muscles that work the legs. The carapace is a hard shield on top of the prosoma, best seen on the moulted skin. When the spider crawls out of its old skin it literally flips the lid off the top of the prosoma and like some weird eight-legged slow motion jack-in-the-box crawls outwards and upwards. Look at the underside of the carapace and you'll see all the bumps and ridges where the muscles attach. Hold the 'lid' up to the light and you'll be able to make out the spider's eyes.

The **abdomen** or **opisthoma** is softer than the rest of the spider's body and you can think of it as a living bag, containing various organs – the heart, breathing apparatus, reproductive system, guts and the all-important silk glands, all sloshing about in blood. The **fovea** is a dimple in the middle of some spiders' carapaces, to which the stomach muscles attach.

Spiders usually have eight **eyes** but there may be fewer in some species. The position of the eyes is important in the complicated business of spider identification.

Twinkling eyes

No matter where you go in the world after dark, as long as it is warm enough for spider life you'll find spider lights. By this I'm referring to the blue, twinkling, and sometimes rather eerie-looking eyes of hunting spiders. Cast a torch beam (hold it near the plane of your eyes – a head torch is perfect for this) around in the vegetation, on the forest floor, even on tree trunks, and you will soon see these little points of light twinkling and shining your torch light back at you. In tropical forests – some of the most spider-saturated habitats on the planet – the scene can look like all the constellations have come crashing to earth, and some fields full of hunting spiders will look like a million sparkling diamonds as thousands of sets of eight eyes reflect your torchlight.

In the same way that the eyes of many nocturnal animals like cats and Foxes 'shine' in car headlights, spiders' eyes reflect light due to a reflective layer, called a tapetum, behind the retina. In animals with a tapetum the light is bounced back through the retina a second time, potentially doubling the sensitivity of the eye to light. This is ideal for an animal that hunts in the dark, although strangely arachnologists cannot really agree whether or not spiders can see well at night, even though their tapetums seem to work in the same way as in nocturnal mammals!

The **chelicerae** are the jaws, used for feeding and defence, situated either side of the mouth. Each has a tiny hole in the tip, through which venom is injected. Some of the big theraphosid spiders, and money spiders, even make noises with them. They are used as cutting edges and filters, helping the spider crush and strain a nutritional liquid soup from its prey. The **mouth** is a small opening on the underside. It can be obscured by various bristles and knobbles on the bases of the pedipalps and chelicerae.

Everyone knows that spiders have eight **legs**. They are full of blood under high pressure and muscles pull the legs as they walk – the high-pressure blood system is demonstrated indirectly when you see a dead spider. Because of a loss of fluid dead spiders curl their legs up under their body as there is no blood to keep them extended. Spiders' legs also contain extensions of their gut! The **pedipalps** (or just palps) are leg-like appendages that sit either side of the mouth and point forward. They are one segment shorter than a leg and are not used in walking, but for wrapping, grabbing, tasting and wrestling prey, and in males they play a role in the mating game.

Sensory hairs cover the entire body, but are particularly numerous on the legs and palps. These are sensitive to vibration, while some are chemoreceptors; they allow the spider to feel and taste its way around.

Most spiders have six **spinnerets**, structures unique to spiders. Each has a cluster of smaller pores called spigots, through which silk is produced.

Some spiders and scorpions have **book lungs**, 'books' of thin 'pages' through which gases are exchanged. Some have spiracles, breathing pores a bit like those found in insects.

Spider semaphore

One of the easiest ways to tell the sex of an adult spider is to look at the tips of its palps. A male looks as if it is wearing boxing gloves! These large bulbous tips to the palps are secondary sexual organs and are used to inject sperm into the female. Some wolf spiders and jumping spiders use them like flags to signal a warning to a rival, or a chat-up line to a female. Her response indicates whether she fancies him. If not he could potentially end up as dinner.

The actual shape and design of the palp tips are species-specific. This is very useful for the arachnologist when making an identification. The palp tips are quite ornate in some species.

During courtship a male Spotted Wolf Spider will signal to a female (here off left of the photo) with his black palps.

Why are spiders hairy?

It's often a reason given for loathing them but the 'hairy spider' is blessed with much more than a fancy fur coat designed to strike fear into the heart of the arachnophobe. It's hair that enables spiders to perform the famous antics of crawling up the smooth surfaces of walls and taking a stroll across the ceiling. Each foot is equipped with specialised hairs called scopulae; scopa means 'broom' in Latin and if you look at these hairs under a microscope you'll see why. Each is divided into hundreds of even smaller hairs, each with a flattened spade-like tip. Each 'spade' uses the surface tension of the microscopic film of water found on most objects. A crab spider has

Each of the individual hairs on this Curly Hair Tarantula have a different job to do – from camouflage and protection to hearing, tasting and smelling.

relatively few scopulae – only 30 – but it can still achieve 160,000 separate points of contact on a surface. This explains why some species with even more scopulae than this can support more than ten times their own weight and walk upside down on the ceiling.

Scrutinise hairs elsewhere on the spider and you'll notice other kinds. The short hairs that cover the body in dense velvet give spiders their camouflage, patterns and colours, which act to break up their outline and help them blend into their background. Larger, obvious hairs are often sensory in function. Some are used as triggers to tell a spider that something is moving; try tickling one of these hairs of a spider with a pin and watch how it reacts. Much finer, feathery hairs called trichobothria are easily 'blown' by even a tiny vibration or breeze, and can literally tell a spider where its next meal is coming from, as they act as ears.

Other hairs perform very specific tasks and are found on some species but not others. Some are used to comb the silk to produce the right texture, and yet others function as hairbrushes to keep all the other hairs in good working order. Water spiders rely on the shape of hairs and bristles on their body to trap a layer of air in place, allowing them to breathe underwater. Adult wolf spiders have special hairs with knob-like tips, which act as handles for their young to grip!

Feed the zebras

The tiny, hyperactive Zebra Spider (*Salticus scenicus*) can often be found on hot sunny walls or windowsills. This spider makes up for its size by sheer 'personality'. If ever there was a spider that could be considered cute, this one is it! It has the standard set of eight but has a pair of particularly big, bright and beady eyes that take up most of the front of its 'face', giving exceptionally good eyesight for a spider. It can apparently detect movement up to 30cm (12in) away. It's a hunter that actively stalks and jumps on prey; its only everyday use of silk is as a safety line which it trails behind itself as it hunts.

Those eyes, combined with a very flexible and mobile body, make this spider particularly endearing as it will spin around to face you and even peer up at you as you look down on it. You can give one a sight test by feeding it small flies or other insects on the end of forceps – they usually respond to tiny movements at around 20–30cm (8–12in) in front of them. Another test is the reaction of a male to its reflection in a mirror – if you get one on a good day it will threaten its own reflection as if it was a rival male. Many of the other jumping spiders (family Salticidae) are flamboyantly decorated with beautiful metallic colours and patterns, especially in the tropics, maybe because they use their excellent eyesight for species recognition.

Zebra Jumping Spider (*Salticus scenicus*) adult, feeding on fly prey, Leicestershire, England.

Wolves of the meadow

Walk through long grass in summer and you may notice small spiders scatter from your path in huge numbers. Get down on your hands and knees and look at these spiders and you'll see a miracle in micro-mothering! These are the wolf spiders. They pursue their prey like wolves, using their big eyes to spot prey before they chase it down and pounce. They do not use a web, and they certainly do not hunt in packs although sometimes it may appear that way because there are so many of them! Because these spiders are nomadic, the females have to carry their eggs around with them in a silken cocoon slung beneath their abdomen – this off-white pellet of silk makes a female stand out during the height of summer. You will probably notice the egg sac before you see the spider carrying it.

The white ball is an egg sac and is carried around by the female wolf spider until the eggs inside hatch.

Now is the time to start spider-watching in earnest (or you could keep one in a vivarium). You'll notice that in due course the female appears to change shape and look a little different – peer a little closer and you'll see that she is now carrying a bundle of spiderlings on her back. Wolf spider mums carry their young for about a week, giving them some protection from predators, a head start before they wander off into the grassy jungle on their own.

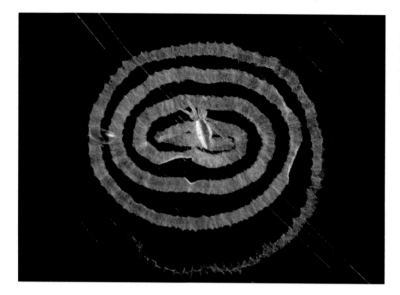

Silk craft reaches its very best in the webs of the orb web-weaving spiders – this species from Borneo has added a spiral called a stabilimentum, the purpose of which is debatable. An attraction to flying insects? Or a flag to stop birds flying through the web and damaging it?

'Silkology': spider silk

Imagine being able to produce a substance with which you could build a home, support your body weight, glue things together, catch your food, use to fly, and that you can roll up and recycle! Sounds like the ideal super-substance. Well, it is, and silk is probably the main reason for the incredible success of spiders as life forms. It has been said that silk is to spiders what flight is to insects and warm blood is to mammals. Even though others such as caterpillars, wasps, ants and caddisfly larvae use it, the spider is the true master.

Watch a spider spinning silk and you'll see, if the light is right, that the silk emanates from the tip of its abdomen. In some species, if you look closer you'll see tiny little projections that the animal seems to wiggle and wave – these are the spinnerets. On the tips of these are microscopic nozzles that look for all the world like little cake-icing implements – these are the spigots from which the silk flows. I use the word 'flow' deliberately – the silk is not forcefully ejected as a certain web-spinning superhero would have you believe. There are no muscles associated with the silk glands. The silk is a liquid soup of protein in the glands in the spider's abdomen and it remains so until contact with the air, when it hardens and stretches. The silk is drawn from the body by movement of the spider or an external source such as the wind.

This can be observed if you watch a spider descending from the ceiling on a line. Watch what it does with its legs. Like an abseiler, it controls the rope as it descends. It keeps one leg on the line all the time, controlling the rate of descent by pulling the silk out of the spigots. There are up to eight different kinds of silk produced by the masters of silkology, the orb web spiders. Different silks are used for different jobs; the silk used to create a web spoke is very different to the silk used to make the sticky bit of the web. The more primitive the spider the fewer types of silk gland and the less advanced its silk technology.

Next time you see a spider on its line, wait and watch if it is climbing back up its own silk – you will see it winding in the thread as it goes.

The doily of death

There are nearly as many different designs of web as there are different types of spider, ranging from simple sheets and tangles of silk, the bane of the houseproud, to the full-blown masterpiece, the orb web. I have had many great moments of total amazement with spiders, but one that summed up the amazing qualities of silk was an encounter I had in Mexico. I had been out hunting for tarantulas on a nocturnal ramble, and on the way back to camp had decided to turn off my head torch to conserve battery power. I was stumbling along when I found myself caught in what I knew was spider silk. Upon turning on my torch, I was confronted by a wonderful tangle of golden threads all around me, and the spinner of this glorious structure was dangling inches from the tip of my nose. It was a member of the *Nephila* genus of golden orb web spiders, spinners of some of the strongest of all spider silk creations. A real beauty of a beast, this female was a big spider, some 8cm across her leg span. Now, here I was, all tangled up like a scene from a horror movie. Then I noticed another animal in a similar predicament! A lizard had also fallen foul of this trap and was suspended in the silk next to me.

The golden bits of the web of a golden orb web spider are actually caused by tiny beads of sticky liquid silk that's purpose is to ensnare insects, and in some cases lizards, frogs and even small birds!

Webs from related spiders are apparently used by indigenous people in South America and New Guinea as fishing nets. The silk is collected in a hoop of vegetation and then lowered into a

Surfing the web

I remember first observing this odd form of spider locomotion by a pool in the south of France. I was aware of small spiders landing on me. I counted 20 or so in 15 minutes. How did they get there and where were they coming from? Then I saw it for myself. One of these adventurous arachnids was climbing my bent leg to my knee. Here it paused, stuck its bottom in the air and waited for a few moments; just enough time for me to see a line of silk. Then the little guy let go and floated away into the blue yonder.

Believe it or not, spiders have been found out at sea and in the air as high as 3,000 metres (10,000ft)! Here they form part of a strange world of aerial 'plankton' that drift around on air currents. Unlike a lot of their windsurfing insect colleagues they lack wings – but they can effectively fly, through a behaviour known as 'ballooning'.

You can witness this for yourself on a dewy morning. Just about any long grass will be laced with moisture-laden gossamer bunting, linking the grass blades. Investigate closely and you may find the culprits, lots of tiny 'money' spiders, either spiderlings or tiny adults. They belong to the family Linyphiidae, and are masters of this mode of transportation. If you collect one of these minuscule animals on your fingertip, and hold it up to the breeze or gently blow on it, you may persuade it to 'balloon'.

If you are lucky the spider will raise itself up on tiptoes and allow the breeze to pull out a thread of silk, which will snake up on the wind. When this develops enough lift to overcome the micro-spider's weight, it lets go and drifts off. A lot of these risk-takers never make it to suitable habitat; millions must perish by landing at sea or in other spider-hostile situations. However, the ability to travel this way means that spiders are great pioneers and are among the first animals to turn up on newly formed islands, sometimes before suitable prey has even become established.

Silk helps many small spiders 'fly' by a process called 'ballooning'.

stream. Any holes created by struggling fish are simply darned up with a needle, just as a fisherman would repair a regular net.

A hectare of field in southern England in the autumn may contain 5.5 million orb webs, wafting in the breeze and covered in baubles of dew. But what you have before you is not a hand-woven swatch of Honiton lace but an efficient and recyclable death trap, just as lethal as that of any tropical species.

Early in the morning before the dew has evaporated is the best time to locate spiders of hedgerow and grassland. The moisture-laden threads glitter with beads of backlit dew.

The orb web is probably the most advanced arachnid architecture, made from minimal materials; less than 500mg (0.017oz) of silk is used for an average web, although if you unwound it by hand it may well exceed 20 metres (66ft) in length! Bearing in mind each web is spun blind, using touch only, and only takes about 20–30 minutes you really do have a miracle of silkology. If dew is in short supply, use a plant mister to create your own at any time of the day, showing webs up to their best effect.

Catch the catcher: extracting spiders from their holes

Many spiders, including some of the world's largest and most spectacular, the so-called bird-eating spiders or tarantulas, live in silk-lined burrows, rarely leaving them except to find a mate or for a split second to grab a passing snack. Look on dry banks and especially walls with crumbly mortar and you'll probably see their doorsteps
– each a ring of woven silk with radiating spokes.

It's a nice hole – but how do you get to meet the owner? Well there is a trick you can use.

Spiders that live in such holes are notoriously difficult to observe and hence identify and study. The obvious solution is to dig them out but this is a very invasive thing to do, as you'll have destroyed its home and releasing the animal back where it belongs is much more difficult! Digging should really be a last-ditch option; but if you trick your spider and utilise some natural behaviours you shouldn't need to go to these extremes.

Flooding them out You can observe these spiders from a distance at night when they sit in the mouths of their holes and await unsuspecting prey to come bumbling along, but at the slightest disturbance they will pick up the vibrations and rapidly retreat to the safety of their lair. Also, for positive identification in the field capture for closer scrutiny may be necessary. This technique works pretty well for most hole-dwelling spiders on level ground and even though it seems pretty drastic the spider suffers little and its home remains intact. Simply pour fresh water into the burrow entrance; the amount of water needed varies depending on the size and depth of the hole. Assuming there is a spider in residence (and not

Out pops a wolf spider! And when you've fished looking at her, she'll be able to return to her undamaged if a little damp home.

tucked away in an airlock somewhere) it will rapidly surface and even leave its burrow. Now block its retreat and quickly place a cup over the animal to secure it for identification. Then when you have finished observing, you can pop it back home.

Tickling and taunting Another way to get them to answer the door is by giving them a buzz and tricking them into thinking you are dinner. There are a couple of ways of doing this. You need a tuning fork (available from shops that sell musical instruments). Strike the fork and gently touch it to one of the silk spokes at the hole's entrance. The spider will come shooting out to investigate. Some attack with such gusto that you can even hear a 'plink' as the fangs strike the metal! Certain keys of tuning fork seem to work better than others – I have had success with C, F and E – as different-sized insects may struggle at different frequencies. Alternatively, try a party blower with a piece of grass taped to the end. Unravel the tube, hold the grass against one of the threads and blow. Another method is tickling the web with a blade of grass, although this doesn't always fool the spider. Scorpions can also be caught using a modification of this technique. Once you have located a scorpion's lair, gently poke a twig or grass blade into the hole and jiggle it around. If the scorpion is hungry it will grab on and in all its stubbornness will not let go. If you wish to fully extract the animal, use a pair of 'soft' forceps (long metal tweezers with foam rubber or even a pad of parcel tape wound around the tips) to stop the animal retreating once it has realised its error.

Some burrow-dwelling spiders such as this Red Rump Tarantula from Belize can be tickled out of their homes by pretending to be dinner.

Picking up arachnids

To get an idea of how potent the sting of a scorpion is, check the size of its chelicerae. Those with larger pincers usually pack their punch this way, and although they may arch their sting and look the part they are probably bluffing. They will give you a nasty nip (one that shouldn't be underestimated either; the big 'fist' contains muscles that operate the pinching action) but their sting is unlikely to do you any damage unless you are sensitive to their venom.

Many of the world's highly toxic species display the exact opposite; they have small, weak and narrow chelicerae, but much more powerful stings. However, having said all this it is always prudent to treat all scorpions gently and with respect, since you can't be sure whether or not the venom is going to kill you, especially given the unknown quantity of potential allergic reactions to venom. Avoid handling them with bare hands if possible.

So how should you manipulate scorpions, or any stinging or biting invertebrate, and pick them up safely, minimising risk to both yourself and the animal? The first thing is a general rule with any rapidly moving scuttling thing; do not act hastily and make a grab for it. This is when either party can easily get damaged or hurt. Instead, try to stop the subject in its tracks, to give you time to think. Easier said than done, I know, especially when you have just flipped over a piece of wood and a small scorpion or spider is sitting there for a split second before there's a panic as the arachnid tries to run for safety while you try desperately to stop it! If you're with a companion not in the know, remember to shout instructions to them too. Once I was being filmed at close range as I flipped a rotten old tree trunk in the rainforest of Queensland. A centipede at least 20cm (8in) long ran out from under it. All was going well until the animal, frustrated at being fielded by me, took the soft option and disappeared up the trouser-leg of the startled cameraman!

If the animal you want to get a look at is sitting in the mouth of its lair or burrow, then it is at a major advantage. It's best to wait for it to exit the hole enough for you to block its retreat. If you disturb it and it retreats, leave it alone and come back again an hour or so later.

The simplest technique – where the terrain allows – is to pop a cup or similar container over the animal. Nets can be useful for this, as are hats and neck scarves; they may not hold for long but give you time to think. Now, depending on species, you can proceed in various ways. If you are on level ground, slip a piece of cardboard under the cup to trap your quarry, or gently coerce the animal to walk into a specimen pot. With scorpions I use a pair of 'soft' forceps if

Gently does it, this Red Rump Tarantula is persuaded to walk onto a hand.

I have them. I have seen a pair of these hastily made from the bent spine of a palm leaf with a twig jammed and bound in the 'V' – these were made for me in Morocco by a local on discovering I had lost my own. I was on the trail of *Androctonus mauritanicus*, one of the particularly potent fat-tailed scorpions, and I have to say this handmade pair, now several years old, are still doing me well. The way the forceps are used is critical too. The best and safest place to take your gentle grip is either side of its prosoma or sting and once the animal is secure quickly lift. Make this unsupported moment as brief as possible, because scorpions do not like being picked up this way; they may struggle and damage themselves or simply turn around and walk up your tweezers! Place it in a collecting vessel for viewing.

If you need to move or handle a scorpion, such as this, may I suggest caution and a pair of padded forceps and not fingers!

If the quarry is a large theraphosid spider, your best bet is to gently convince the animal to step into a collecting tin, tub or pot. Even a slow spider can turn into an eight-legged rocket when it perceives its life is in grave danger. Spiders are particularly fragile – that abdomen is like a large water-filled balloon and must be treated as so – but with large spiders and when the handler knows what he or she is doing it is possible to pick the animal up by hand, gently but firmly (and with much confidence) gripping the animal either side of the prosoma. A word of warning – if you are not confident or sure of yourself, or the spider, do not do it!

It's worth mentioning the two main defence strategies of these animals. Their preferred one depends on which continent you are on. American theraphosids are not usually that aggressive with their 'fangs' but instead kick nasty urticating hairs at their attacker. Under the microscope these hairs resemble a collection of barbed and spiked Zulu spears, and if they get in your eyes, nose, mouth or onto soft skin, they really itch and make you sore. If you are super-sensitive to them, like me, they can ruin your day. Learning which species tend to 'kick up' like this comes with experience so treat all with the potential of doing this. The rest of the world's tarantulas lack this ability and instead defend themselves with their 'fangs'. Just avoid these sharp bits; they hurt!

All spiders have mouthparts like these exhibited by the King Baboon Spider; these are the chelicerae or fangs and are two good reasons why you need to be careful when handling large, aggressive or venomous species.

The black light phenomenon

Nobody really knows the exact purpose of this rather handy phenomenon, but it is the most useful little trick when the naturalist is out at night looking for scorpions. Even with the aid of a torch they can be incredibly hard to spot. This is where a 'black light' (particularly effective are those that produce light at 395nm) comes in very useful. These are now readily available in small and easily portable forms. Buy one as powerful as you can. When out at night scorpion-hunting, wave this light about in front of you as you go and pretty much every scorpion that you come across will take on an unearthly quality, fluorescing a weird bright green, making them very easy to spot and watch.

Nobody really understands the function of this fluorescing but most scorpions, like this Emperor Scorpion from West Africa, do it under ultraviolet light.

~ FINDERS KEEPERS ~

Those spectacular and ephemeral masterpieces that are the orb webs can be preserved and kept. The process is a little fiddly but can be well worth the effort as part of a spider study; it is also a great way to measure and compare the structures of the webs of different species.

You will need:

- spray paint
- artist's fixative or hairspray
- newspaper
- coloured card
- scissors

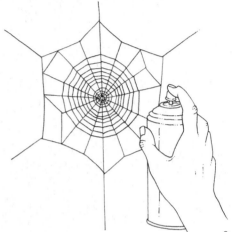

Step 2

1. Choose a nice still day and find your web. Make sure it is free of dewdrops and check its maker isn't in residence (look around the edges of the web, especially in curled-up leaves). The spider will create another web once you have left.

2. Take a can of spray paint – white or black are good – and holding a sheet of newspaper behind the web to stop you getting paint all over the surrounding vegetation, spray the web evenly and lightly on both sides from a distance of about 40cm (16in). Too close and you'll blow a hole in your web.

3. Leave it to dry for a while, then repeat.

4. Now spray both sides of the web with artist's fixative (available from art and crafts shops) or hairspray, to make it sticky.

5. Before it dries take a bit of card, big enough for your web to fit on, and of a contrasting colour to that with which you sprayed the web. Carefully line up the card with the web and push the card onto the silk so that it sticks in the right place. Once the web has touched the card you cannot move it without ending up in a right messy tangle! If you've done it right you should have a perfect web on the card.

6. Use scissors to cut the supporting strands, and give it another coat of fixative to make sure it's held in place.

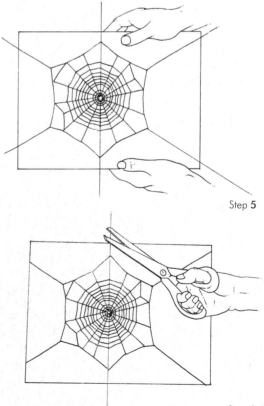

Step 5

Step 6

Air crabs: harvestmen

A wonderful, understated and often overlooked group of arachnids – the harvestmen are worth a closer look.

Look under ivy that has grown against walls, or among ferns and thick vegetation, in dark corners of the shed and under windowsills, and you are sure to turn up a gangly life form that when disturbed wobbles off in a manic random sprint that is somehow better co-ordinated than its cotton thin legs should allow! The Brazilian name for them, which I rather like, is 'giro mundo' which refers to their speed and agility. We know them as harvestmen, as the large adults appear in late summer.

The technical name for them as a taxonomic order is Opiliones. They differ from spiders, which they superficially resemble, by not having venom glands, not being able to produce silk, and by having their body parts fused together to form a single button-like body, usually suspended between the long vegetation-spanning legs when the animal is active.

They may not seem all that charming but they are definitely worth a closer look. Carefully catch one, avoiding making it shed its legs (this self-amputation is a defence strategy). Have a closer look and you'll see a pair of simple eyes sitting either side of a turret on top. Look underneath and you'll see the mouth – a tiny spot. To the front is a pair of tiny crab-like pincers or chelicerae, used to crush and masticate their food of carrion, rotting vegetation and small insects. The males can be told from the females by the long spurs attached to their chelicerae, which stick out like a pair of horns. After handling it, smell your fingers! Harvestmen have a pair of defensive repugnatorial glands that emit a cocktail of unpleasant and irritant quinones, to discourage small predators.

If you have any doubts about identification, look at the body. A harvestman's is an all-in-one affair, no two halves here!

Monster on a pinhead – pseudoscorpions

One of my favourite invertebrates in the whole of the garden is the pseudoscorpion – a great-looking beast and one closer to home than you might think. I'm willing to bet that you have them in your own garden, the school playground or park hedge, just about anywhere there are a few leaves. My best catch yet was 19 from a little pile the gardeners missed in the middle of the BBC car park! These are not scorpions, they have no sting, are only 2–5mm (^{1}/$_{12}$–^{3}/$_{16}$in) long and are completely harmless.

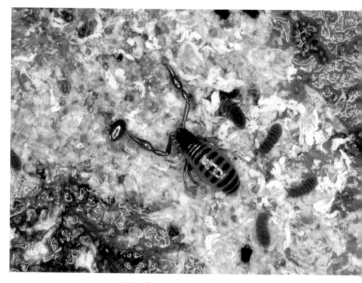

It's often a surprise to most people when they corerum realise these little scorpion-like life forms are all around them. This is the Littoral False Scorpion which lives on the coast, often sheltering from the tides bye sitting in bubbles trapped in barnacle shells.

Their prey includes any small creatures they can get within their venomous grasp, such as mites, nematodes and springtails. It overpowers these with a lethal injection given through the tips of its hollow pincers. There are some 25 species living in the UK and they can be found anywhere from old barnacle shells on the seashore to the debris in your garden, even among dusty papers in your home, where they feed on booklice.

Personal fauna

Another little arachnid that you may end up culturing unwittingly is the hard-tick (named after the hard shield-like scutum on top of its body). These parasitic creatures have a rather interesting life cycle and are fascinating in their own right but even the most curious and open-minded naturalist would be hard-pressed to study them while they are attached to his or her person, particularly as the sorts of spots the ticks like to sink their jaws into are soft, moist, thin-skinned and often sensitive areas! It's worth being aware of them as they can transmit a few really nasty diseases, such as Lyme disease, and tick fever in Africa.

When hard-ticks reach the blood-requiring stage of their life cycle they do what is known as 'questing'. They crawl to the tips of vegetation and await a host, with front legs extended. They know when one is coming due to the very sensitive 'Haller's organs' on the tips of both front legs. These are sensitive to movement, humidity, and maybe even chemicals associated with a host's body. If the host brushes the vegetation the tick is on, it will inevitably pick it up ticks

are masters of hanging on. With their grappling-hook limbs they hike to a soft spot before scissoring through the skin using their barbed mouthparts (hypostome), which they jam into the wound. This is why a tick is very hard to remove.

Well known and unloved, because of their bloodsucking habits, ticks have a fascinating life cycle.

Remove them we often must, but how? You'll hear lots of rather ludicrous solutions, but the best and quickest way is by simply grasping the tick between your fingernails and pulling with a twisting action if it's big enough to grip, or if it's small then scrape it off with a fingernail, making sure you have not left any bits of tick in the wound. Dab the site with antiseptic, and keep an eye on the bite area for secondary infection or any reaction. Seek medical advice if any swelling or redness persists. Removing ticks can become quite an evening pursuit; armed with a mirror and torch I have removed more than 200 from various sensitive regions of my person after a day in the field in Africa.

In areas with gatherings of large mammals, such as deer or antelopes, you can expect ticks. They like to 'quest' in vegetation that is not too dry, so long grass and damp places such as water holes and rivers are a good spot to pick them up. You can reduce the risk by tucking your trousers in your socks and your shirt into your trousers, and by wearing light colours so you can see the ticks as they crawl around looking for somewhere to plug in.

Crustaceans

Familiar to anyone who has gone poking around in rockpools and crevices down by the shore, crustaceans are among the first creatures a young proto-naturalist will learn to find, catch and stuff in a bucket. Some are sessile, others active, we eat them and they are commercially important. You also sometimes find them wandering around on your carpet at home, and most gardens have a population of them.

This is primarily an aquatic group, but has also taken on the insects on dry land. The most famous land crustaceans are the woodlice or pill bugs, with life cycles that cut ties with water, but land crabs and some hermit crabs still have to return to the watery realm to breed. From the fully aquatic crabs, lobsters and crayfish through to the woodlouse and slater, the naturalist can find the entire story of the invasion of land by these arthropods. Flip over even the most land-based of them, the woodlouse, and you'll see what I mean. In an evolutionary sense woodlice have really only just stepped onto the land; most of their isopod relatives still live in water. If you catch one, and gently turn it over, as well as seven pairs of walking legs, you'll see a small pale patch towards the back end. This is actually a set of gills, and contains the secret to their conquering of dry land.

Water Hog Louse – an aquatic close cousin of the Woodlouse.

Put simply, gills are structures designed to absorb oxygen from the water into the blood, and to get rid of waste gases. To work efficiently, it helps if the skin covering the gills is thin so blood flows as close to the surface as possible. This is why gills in other animals are often transparent, and in animals with red blood such as fish and tadpoles they look red. They will often be flapped about to speed the flow of water over them.

Those crustaceans that live in water have exposed gills, like flattened legs folded back at the knee. Prawns, freshwater shrimps and that wet woodlouse, the Water Hog Louse (*Asellus aquaticus*), all have them; just look under their rear end (the true crabs have gills but they are hidden within a cavity in their body) and you'll see them in action.

Woodlice have these structures too – they are just not flapping about all over the place. Use a magnifying lens of at least 15× and you'll see water sloshing about between the plates; they might have invaded the land but they are still using gills like their aquatic brothers. These gills must stay wet to work properly, and are the reason woodlice are only found in damp places – they still need water to breathe!

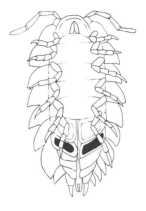

Woodlouse gills.

Crabs' secret gills

Next time you lift a crab out of water watch what it does; it will probably start blowing bubbles from around its mouth. This is your clue to how it breathes. Imagine the crab is a lobster, but with its 'tail' folded up underneath between the legs, and the carapace (the shell on the top of its body) expanded to cover the gills, which are held in cavities inside.

The following demonstration may help you understand. The bubble-blowing is how the crab ventilates its gills. The mouthparts waft water into the cavities either side of the mouth, under the carapace, over the gills, and it 'leaks' out through pores between the legs. If you use a pipette to squirt food dye in front of a living crab, you'll see how the liquid gets sucked in and where it comes out.

A good way to get to know your crustaceans is to find the empty, freshly moulted exoskeleton of a crab. Here you can see all its features inside and out including the finger-like, usually hidden gills (the dark brown bits).

Catching your crustaceans

'Crabbing' usually involves catching as many of the unfortunate animals as possible and stuffing them in a bucket before the tide turns and ruins the fun. One way to find members of this diverse group is to simply turn over stones and seaweed to expose those underneath; this is still an excellent way of doing things and you'll find many other

Leg action

Find a stone with a colony of live barnacles on it (dead ones have usually lost their 'front door' valves, leaving a gaping hole). Pop your colony in a tank of saltwater and watch carefully; you'll see the doors open and a pair of feathery organs will start to flick out. These are the barnacle's limbs (imagine these animals a little like crabs that have settled down upside down with their legs sticking up). Their feathery structure gives a large surface area, for sieving particles out of the water. Many aquatic animals use this form of feeding, but the barnacle is pretty proactive at getting its dinner. Does the rate of 'feeding flicks' change if you alter the water temperature? You can also try and watch the feeding itself if you squirt little bits of chopped-up and liquidised food in the water using a pipette.

crustaceans such as barnacles and slaters as well as a whole plethora of other rockpool creatures. Just remember the golden rule of putting things back where you found them.

An old shaving mirror tied to a stick is useful for looking underneath ledges and in cracks and crevices; the mirror also illuminates these dark places by reflecting sunlight. A stout pole is also useful. It acts as a handy third leg on the slippery weedy rocks, and you can use it to probe likely-looking crevices and lairs. Usually if a lobster is in residence it will use its pincers to defend its corner and grab on. The lazy or the smart will try using bait. This technique works for everyone, from someone dangling a crab line over the end of a pier to fishermen using crab pots. Many free-moving crustaceans such as crabs and lobsters are scavengers and are attracted to dead animals, which they smell over considerable distances.

One of the most fun and satisfying seaside pastimes. Rockpooling, the best way to meet a crab.

Many happy hours of my summer holiday as a kid were spent gazing into rockpools. The pretence was that I was catching prawns for dinner. I would pre-bait rockpools with meat of any kind, from scraps begged from the butcher or fishmonger to the contents of my ham sandwich. These baits attracted many scavengers, including crustaceans of all kinds. Crabs would come scuttling from all corners, some even skipping over from neighbouring rockpools.

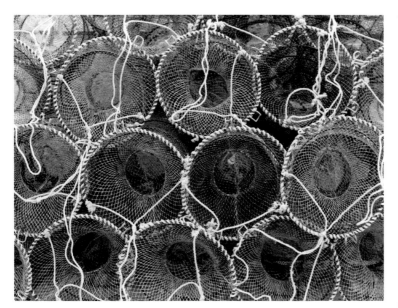

The same principle used in lobster pots by commercial fisherman can be utilised in miniature pots by the naturalist.

They would congregate around the bait like hyenas around a kill, joined by smaller hermit crabs and prawns. This was my first lesson in just how sensitive and effective they are as scavengers.

Fishermen set traps to exploit this; mesh pots and creels are baited and left. Lured by the irresistible scent of the food within, crabs or lobsters walk around until they find the conical one-way entrance. The naturalist can harness the same principle. I have used baited hollow bamboo stems or logs in Australia to catch freshwater crayfish or 'yabbies', and even wooden boxes crudely nailed together to the rough specifications of a lobster pot.

A scaled-down version of the above is a plastic drink bottle, cut up, reassembled and weighted down. It will work for smaller shrimps and crabs, or you could try a net, especially when working in open water. Larger prawns and shrimps can be caught using regular nets and when working on soft substrate hunting for burying species such as the true shrimps, the best bet is a flat-bottomed net that can be scooted through the surface of the sand. These are usually available from a specialist net maker or natural history supplier.

For the free-swimming and larval stages try plankton nets; these are long tapered nets that end with a collection vessel at the end and can be towed slowly behind a boat. They are often very effective and you'll be surprised at how many shrimp- and prawn-like creatures are floating about out there in the open ocean. All the marine crustaceans have a larval stage and are part of the great drifting masses that help to feed the rest of the food chain.

Shrimp net

House swap

The familiar hermit crabs are not true crabs but somewhere between these and a group of crustaceans called squat lobsters. Stare into a rockpool for long enough and eventually a usually sedate mollusc will get up and sprint off!

The habit of living in old shells probably started when a distant relative started living in a crevice, making forays into the open to snatch food; from this it is only a short hop to the mobile hideaway. Hermit crabs went on to scavenge their way around the shallow seas and beaches of the world, the borrowed shell giving them excellent protection from both predators and the elements.

This land hermit crab is well protected in its shell. However, in really hot weather it will dig into the sand or seek shade in the bushes.

As a hermit crab grows, its accommodation can get a little tight, and eventually it needs to move up the property ladder. The process of moving house is near impossible to witness in the wild, but they take well to a short stay in a marine aquarium, where you might get lucky enough to witness it.

Collect a variety of different-sized gastropod shells and a hermit crab. The tricky bit is persuading your crab to leave its house. Many can be extracted by gently pulling and maintaining the pressure on the front limbs, which is a little bit like a tug-of-war as the crab grips the inside of the shell with special modified limbs. Holding the shell and blowing on them is also a trick that works well with land-based species.

Coconut Crabs, the biggest land-based crustaceans, are just a big hermit crab – they start off life living in shells but when they get full grown finding shells to fit becomes much more tricky.

A naked hermit crab is a strange sight; the back end of its body is a floppy and vulnerable sack and you can also see the limbs that have been turned into grippers. Now place your upset crustacean back into your tank among a selection of shells including its old one, and watch what happens. The crab quickly scuttles around, measuring the shells for size with its legs and using the large pincer like a calliper. When it selects its 'des res' it rapidly reverses into its new home, panic over. In the wild the actual change-over is equally rapid and nervously performed but usually the animal will

Common Hermit Crab adult, with Hermit Crab Fur Hydroid on its shell.

have pre-selected a shell and the swap is very much like a bashful person changing into their swimming trunks.

Some hermit crabs have a symbiotic relationship with anemones, which they carry around on their shell for camouflage and protection. When the shell is swapped the anemone is also transferred delicately, and a crab performing this loving act looks for all the world like a proud gardener re-potting a prize begonia.

Brine shrimps and triops

Some remarkable crustaceans can colonise temporary pools of water. They do this by turning around their life cycle extremely quickly, and sitting out the extreme conditions of drought in the form of a super-tough life support capsule of an egg. These eggs are like living dust; on being exposed to water and rehydrating they'll hatch, grow up and lay more eggs in a few days.

Brine Shrimps are in many ways the perfect pet and a crustacean you can rear at home. Here are the nauplius larvae hatched and hatching.

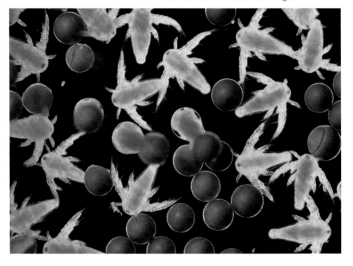

The animals that show this adaptation best are easy to get your hands on. Both brine shrimps and triops can be easily purchased in educational science kits, from biological suppliers and the former also from good aquarist stores where they are often used as live food for fish. Just add water, keep them warm and watch very carefully for the first signs of life.

The barnacle with a dark side is *Sacculina carcini*. The females of this parasitic barnacle prey on Shore Crabs, infecting as a tiny larvae, before ramifying through the tissue of the host, eventually becoming visible to the human eye as a large shiny yellow-brown mass (not to be mistaken for the granular egg masses of the crabs themselves) on the underside of the unfortunate animal. Infected crabs cannot shed their exoskeletons as they would do normally, so the uppersides of infected crabs become colonised with barnacles and tube worms – a sure sign that if you flip the crab over you'll find the parasite in residence.

This photo shows the parasitic barnacle (the yellow) attached to a crab.

Collecting the crusts

You can collect crab exoskeletons easily; they form a useful reference collection and help you learn. A collection of the carapaces of Shore Crabs (*Carcinus maenas*), displayed on card or mounted on strips of double-sided sticky tape, makes for an impressive sight. You'll see a huge range of sizes and variation in colours and patterns, at different stages of their lives (they shed their exoskeletons regularly like all other arthropods) and between individuals.

Walk any strandline on any beach and you'll find evidence of crabs in some form. Often it'll be the odd leg or carapace after storms, but when many animals are moulting a strandline can become a wrecker's yard of empty husks, clues to what life is out there.

It is easy to tell a moulted exoskeleton from a dead crab. Dead crabs get very smelly very quickly. A moulted skin doesn't smell so bad and you can see the spilt behind the carapace, where the crab emerged (it's a bit like flipping the lid on a box and crawling out backwards).

A moulted skin is a perfect replica of the animal that once lived in it. Investigate the old skin carefully and you'll see how the 'tail' or abdomen of the crab has been curled up beneath its body, and inside you'll see the positions of the fluffy gill filaments that are usually hidden inside the protective armour-plating provided by the carapace. When you find a cast like this while out rockpooling it is sometimes worth casting your eye around and turning a few stones as you may

Finding the empty carapaces of rarely seen species such as this Masked Crab give the naturalist clues as to what is living nearby offshore.

find the 'peeler' itself; the name given to the freshly moulted crab that is still soft, pale and defenceless, its new exoskeleton having not hardened up yet.

If you endeavour to keep any crustacean in captivity you may well not only be lucky enough to see the process itself, but you can often retain a complete exoskeleton as a souvenir too.

The moult of a Pennant's Swimming Crab.

Think tanks

Forget your bucket and spade; this is the piece of equipment essential to anyone visiting a beach. As soon as we lift an aquatic animal out of its realm its beauty is instantly compromised; floppy bits that are usually supported by the water just flop. Animals that live in shells shut up shop and little mechanisms like the feeding motions of a barnacle or the breathing and burrowing mechanisms of a prawn are all lost. By tank I mean anything that can hold water and has transparent sides to it. I prefer to use cheap plastic tanks. They are lightweight and don't smash as easily as glass. The only disadvantage is that they can quickly get scratched in the often abrasive environment of the shore.

Anemones are an example of an animal that look their best when viewed in their natural element.

Pop anything living into a tank of water and they relax, feeling like the tide has come in. Sooner or later they will start getting on with what they do best and this time because you have a window into their world you'll be there to observe. Mussels will open up, show you their siphons and any internal residents such as Pea Crabs (*Pinnotheres pisum*), rarely seen alive in any other way, may make their presence known. Blobs of shapeless jelly become strange egg-cases or anemones, a section of seaweed stem is a pipefish and that starfish that looks dead out of water starts gliding around at surprising speed.

It's a world you can only appreciate with a crab's eye view – you can either stick your head in a rockpool or use a little trick with a mirror.

Mirror on a stick

When rockpooling, instead of turning over stone after stone you can make a naturalist's version of a dentist's mirror to allow you to look under overhanging rocks and in crevices. Take a piece of thick but bendable wire and using tough reinforced duct tape (from a DIY shop) attach a travel shaving mirror to the end of it. By bending the wire you can achieve the right angle to look in all those previously inaccessible nooks and crannies and even get an idea of what the world looks like from a crab's point of view. A simple mirror is also an invaluable 'gadget' for looking into holes, cracks and crevices. Use one to bounce sunlight into dark places; it's better than using a torch, although these are useful on dull days too!

The softies: molluscs

The name mollusc is derived from the Latin word 'mollis', meaning 'soft', and, though many are encased in a not-so-soft shell, the soft body is something every member of this group has in common. Incidentally the word 'malacologist', someone who studies molluscs, also means 'soft', though derived from Greek this time. The 10,000 or so species show so much variety that the naturalist will find it hard to believe they all belong to one phylum, Mollusca.

Within Mollusca, the slow and basic limpet is at one end of the scale while at the other end you have the intelligent and active cephalopods – the octopuses, cuttlefish and squid. The latter have abilities that begin to compare with vertebrates. I will never forget the first time I looked into the eye of a live octopus as it slithered from an old drinks can that had been its home onto the deck of a boat. When I popped it in a tank for further study it changed colour from red to ghostly bleached white and appeared to look at me with an eye so soulful I immediately felt guilty and released it! The same couldn't be said for its lowly brethren, the clams that I ate for supper that same day.

As a group, molluscs' huge diversity of form may have something to do with the 600 million years that have passed since these animals first came creeping into being. Since then they have diversified into chitons, clams, nudibranchs, snails, tusk shells, octopuses and squid, and can be found in most ecological systems from the deepest of the ocean's trenches to the tops of mountains, from lakes and streams to barren deserts.

An idea of mollusc diversity from the highly advanced octopus (above), to bivalve razor clams (below) and the classic and typical snail (left).

Characteristics of molluscs

All molluscs have an unsegmented soft body, which must be kept moist to stay alive.

Most have internal or external shells made of calcium. The hard shell in snails and bivalve molluscs is relatively large and protects the animal. In the fast-swimming squid, however, the shell is reduced to a small pen-shaped internal structure, while in the octopuses, nudibranchs and some of the land-living slugs, shells have been lost entirely.

All have a mantle – a lobe(s) or fold in the body wall that lines the shell, and contains glands that secrete the materials used to form the shell.

All have a muscular foot which can be put to many uses: moving around; digging into the sand or mud where they live; or clinging on to hard rocky surfaces so tightly that even the roughest sea cannot shift them. In the cephalopods, the foot has evolved over thousands of years to become many arms or tentacles.

The largest of the bi-valves – the spectacular long-living and sadly rare Giant Clam.

The seven different classes of mollusc are:

1. Aplacophora – marine mud-dwelling worm-like molluscs

2. Polyplacophora – chitons or 'coat of mail shells'

3. Monoplacophora – primitive marine limpet-like molluscs

4. Scaphopoda – tusk or tooth shells

5. Gastropoda – single-shelled snails and shell-less slugs

6. Bivalves – molluscs with two halves to their shells

7. Cephalopoda – squid, cuttlefish and octopuses

Not all molluscs are slow – some, such as the squid, are lightning fast active predators. This one gives us a good look at its tentacles used for capturing its food.

Belly-feet: the gastropods

Slugs and snails belong to the molluscan class known as the gastropods, which means 'belly-foot' in Latin. They all get around on one large, muscular foot, which sits underneath their whole body. The only real difference is that snails have a hard shell they can crawl into while slugs do not. Some species of slugs have a tiny vestigial shell that looks a little like a small toenail, and most other seemingly shell-less species have a little one inside their bodies. And there are a few snails that do not fit inside their own shells!

shell

sensory tentacles

muscular foot

unsegmented soft body

mantle

slime glands all over the body

The skin of a snail or slug is covered in **slime glands**, which secrete a thick mucous or slime. The skin is then irrigated with the stuff; microscopic channels (visible with a hand lens) allow the slime to ooze.

The **mantle** is the thicker, tough, rubbery flesh around the mouth of the shell in a snail and the hump of a slug. It serves many purposes including secreting the shell in snails, and forming a curtain that the snail hides behind when it retreats into the shell. It is also responsible for the screen of bubbles that are blown at any intruder or over-curious naturalist.

The **shell** is the hard protective case of snails. In most cases it is made from a mixture of proteins and polysaccharides called conchiolin that bind together, forming tough, chalky calcium carbonate. The **periostracum**, found on some land snails, is a thin layer of proteins and sugars that give the shell its colours.

The **breathing pore (pneumostome)** of molluscs that have a 'lung' is always found on the right-hand side of the mantle and can be opened and closed, so sometimes you can look for it and not find anything. This hole leads to a chamber called the mantle cavity. This is used as a 'lazy lung' – air wafts in and a network of blood vessels in the skin picks up all the oxygen required.

All land slugs and snails have four **tentacles**. Each has a dark eyespot at the tip. It can't see detail but is very sensitive to light. The animals also use their tentacles to feel their way (you can steer a slug or snail by gently touching its tentacles one side at a time). Each of the four is also covered in taste buds, to taste and smell food. The lower ones are sensitive to food up to 20cm (8in) away, while the larger top pair have been shown to work over 50cm (20in) or more.

It is the shell of a snail that makes it instantly recognisable, even when the animal itself is long dead and gone. It's not just the gastropods that have a protective calcium case; bivalves also have a two-part shell that performs a similar function. Its main purpose is one of protection, both from predators, and also the ever-present threat of desiccation. Slugs do not have this advantage and this is the main reason why you can find snails in drier habitats than slugs. Before you start feeling sorry for slugs, their lack of shell means that they can squish into and hide in smaller, tighter spaces than snails, and can move faster!

Make time for slime

For a land mollusc, body slime is vital. It slows down water loss through the thin skin, and works in the same way as adding gelatine to water – by binding the water molecules together in its matrix to form a jelly. Because slugs do not have a shell they are more vulnerable to desiccation, so they have thicker slime. Having said that, some tough slugs and snails can still lose around 50 per cent of their water and live to rehydrate another day. Gastropods also produce a different, more dilute (97 per cent water) slime – this is the pedal slime produced by glands on the foot. This is what is called a viscoelastic – it is both a glue and a lubricant, allowing the snail to both stick (when the foot muscle is stationary), and slide when a pressure is applied.

When conditions get very cold in winter or extremely dry in the heat of summer, snails can retreat into their shells and seal off the entrance of the shell with a thick mucous, which dries to form a waterproof seal called an epiphragm. You can often see this if you peel a snail off a dry wall in summer.

Wrinkles and rings

In many organisms, growth occurs in seasonal spurts, when conditions are best. Just as with the growth rings in a tree (see page 317) or a fish scale, the same can be seen in the shells of molluscs. Look closely at a shell and you'll notice little ridges and sometimes slightly different colours to sections of the shell. This is explained by the way in which a snail's shell grows.

Shells grow when the mantle adds new material to the mouth/edge of the shell. Snails have good times, where food is plentiful and they grow well, but also hard times – winter, and also really dry spells in the weather when the snail hardly grows at all. Boundaries between these periods are what cause the ridges and wrinkles in the shell. You can see and count these on the surface of some mollusc shells, although they can be a bit bunched up near the hinge or 'umbo' region and in old shells they can be obscured by growths or deformation in the periostracum. Scientists can even identify evidence of the spring and neap tides in species which live between the tides, such as clams – times when the animal was submerged or exposed for the longest period of time and therefore either fed well or went hungry.

The Quahog, while not looking like anything special, is the oldest known living organism. One specimen called Ming was over 500 years old!

Skin-breathing

All slugs and snails can breathe through their skin to some extent, as well as through their breathing pore. Keep a few Great Pond Snails (*Lymnaea stagnalis*) in a jam jar next to your bed at night and you'll actually hear them breathing, a kind of popping noise as they come to the surface and open their pore. Some other water snails have gills and extract oxygen from water that fills their mantle cavity. Others have a combination of both.

Great Pond Snail seen here at the water surface with open breathing pore or spiracle.

~ SLIME-SURFING ~

Both slugs and snails produce a trail of slime or mucous which is often visible as silvery marks on walls and patios. Slugs leave a continuous band while snails leave a line of splodges. It is this slime that allows these animals to get about so efficiently without legs.

You will need:
- small piece of Perspex or clear glass
- a slug or snail
- misting spray of water
- hand lens

Take a snail or slug, place it onto the Perspex or glass and wait for it to start moving. If your gastropod refuses to budge or even shake a tentacle after a few minutes, it can be persuaded to go for it if sprayed gently with a mist spray or water flicked off the bristles of a toothbrush.

When your mollusc is on the move, gently lift up the clear sheet and watch it from below. What you should see is what looks like a conveyer belt of dark and light bands moving along the underside of the snail's foot, called a sole. These are bands of muscles; the dark ones are raised and 'stepping' forwards, while the pale bands are in contact with the surface. But this would still not work if it wasn't for the famous secret ingredient – the pedal slime. Glands on the underside of the snail produce copious quantities of this stuff, which lubricates the ground and allows the snail to almost surf along. It acts as an adhesive and a lubricant, depending on how the muscles of the sole are acting upon it.

Gliding on glass – the alternating ripple of muscles can be easily seen from underneath.

Touching, feeling, smelling, seeing

Gastropods can protect their tentacles by drawing them inside out and sucking them back into the body, just like a glove finger can be turned inside out. Try shining a torch behind the snail then gently tapping a tentacle with your finger. Watch as it rolls in on itself. When the coast is clear, it squeezes blood back into the tentacle and it rolls out again. If tentacles do get damaged it's not the end of the world as the snail simply grows new ones!

Making more molluscs

Most terrestrial molluscs are hermaphrodite – each animal is both male and female. But there are many variations on how this works. Some can reproduce by themselves, while others need to meet another hermaphrodite. In many aquatic molluscs you have separate males and females and just to confuse things some, such as the Slipper Limpet (*Crepidula fornicata*), can change sex. Being hermaphrodite is supposedly a body layout that has evolved to increase the efficiency of reproduction.

The act itself can be witnessed when conditions are right; the perfect weather for love is damp, dark and warm. Take a torch out on a suitable night. Snail and slug mating can be as simple as a kind of mollusc kiss-chase, with one animal picking up the trail of the other and simply following this until the potential mate is found, then after a quick (well, quick for a mollusc!) embrace both animals are on their way, having exchanged sperm via the genital pore on the side of their bodies.

Things reach another level of complexity with certain snails. When a Garden Snail (*Cornu aspersum*) meets a mate head-on, they rear up and with their bodies pressed together they 'kiss' for a while, then to get each other in the mood they fire 'love darts' called gypsobelum. After stabbing each other with these sharp shards of shell-like material they mate. Nobody is really sure what these darts do; they could stimulate the other snail into producing sperm or they could inhibit the other snail from mating with another. Whatever the reason, I bet you never expected to find this kind of passion going on behind the potting shed!

A stack of Slipper Limpets – the animals on the bottom of the pile are females while the males are on top; if the females die then the next in the queue becomes a female!

The surprising love life of a snail is well worth investigation – here two Common Garden Snails are at it, having stabbed each other with javelin like love darts.

It gets even better than this. In fact the Great Grey Slug (*Limax maximus*) and its relatives have some of the most spectacular sex in the animal world and if you are ever fortunate enough to witness it you'll feel like giving them a round of applause when they have finished. When the slugs meet, they run around each other and work up a bit of a lather, producing loads of mucous, then one slug initiates the act by 'going upstairs'. They climb up a vertical surface, still chasing each other. When they find a mutually acceptable spot for their affair they start tickling each other with their tentacles. While they get more and more involved with each other they continue to ooze slime and further entwine, until they make the ultimate lovers' leap! This is more of a lovers' abseil really, as they slowly lower themselves on a rope of slime, which can be as much as a metre long. It is at the end of this rope that the relationship is finally consummated.

Finding eggs

When looking under stones or digging soil, you might stumble upon what look like little clusters of miniature translucent ping-pong balls. These are the eggs of slugs and snails. The size of the clutch laid varies a lot, but anywhere between 10 and 100 is common. They lay them in damp crevices where they will not dry out. Depending on the temperature they will hatch within a few weeks. They are quite fun to collect and watch hatch. You'll see the little molluscs inside the egg just before they actually break free, a miniature molluscan miracle well worth watching through a hand lens or a binocular microscope.

One of the most sensual performances in nature – a pair of Leopard Slugs entwined, dangle on a thread of their own mucous.

If you find the immaculate little spheres of slug or snail eggs while poking around in the garden or wood, have a close look or keep them and hatch them; you'll get to see the developing embryos and you'll witness them hatching.

Pond snail eggs are easily found by flipping over lily pads or looking closely under submerged stones and weed, while marine snails such as periwinkles lay theirs among seaweed and on rock surfaces. The principles are the same – a mass of embryos embedded in a protective jelly coating. Sometimes if you are keeping snails in an aquarium they will lay their eggs on the glass; keep an eye on these and observe them through a hand lens and you can follow their progress.

Other mollusc eggs that you may well come across, especially when rockpooling or walking the strandline, may not appear to be eggs at all at first; look out for the pale yellow spongy egg masses of Common and Red Whelks (*Buccinum undatum* and *B. antiqua*), known as 'sea wash balls'. Each of the pods contains more than 10 eggs, but only one will ever survive to adulthood as the first to hatch eats its developing siblings. An odd delicate sandy structure that looks like a low sand cone with the top knocked off is the camouflaged egg mass of the aptly named necklace shells (*Polinices* spp.). Rubbery clusters of pods, known as 'sea grapes', are the egg masses of cuttlefish.

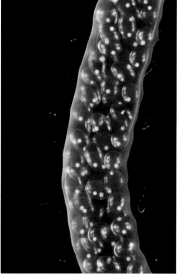

Look under lily pads and on the leaves of pond weed and you'll find these slimy sausages and baubles, which are the egg masses of freshwater snails. Look closer and the next generation are clear to see. This is a close up the eggs of the Great Pond Snail in their tubular egg case.

A common sight on beaches is the sea wash ball which is the egg cluster of the Common Whelk.

Use a torch or mirror to illuminate overhangs and crevices in the intertidal zone of a rocky shore in spring, and you may well find the egg cases of another notorious predatory rockpool mollusc, the Dog Whelk (*Nucella lapillus*). Strange pink and white gelatinous strings found in pools at this time could also be the eggs of various species of nudibranchs. Some look like 'silly string', others resemble frilly garters.

Like bottles lined up on a shelf, each of these 'vases' contains a cluster of Netted Dog Whelk embryos.

Not so soft – defence strategies

Many animals make a meal of a mollusc, everything from voles to bears and thrushes to whales – but if you are feeling sorry for these softies, then be reassured that molluscs are far from defenceless. I'm not just talking about the physical barrier of the shell here either, although having said that many shells come with spines and bristles and a retractable front door called an operculum, making the fortress just a little more impenetrable.

Just pick up and prod a Garden Snail and fairly soon it will produce a thick, green, bubbly ooze – an unpleasant deterrent to any determined mollusc predator. Various slugs when threatened will also produce a super-viscous, foul-tasting, smelly, often coloured defensive mucous from glands on their mantle which, when it gets on fingers, fur or feathers, is quite unpleasant. Garlic Snails produce a strong odour, reminiscent of the familiar and famously odiferous plant bulb.

A dancing slug – this Black Slug has hunched up and is now in the process of defending itself by producing mucous and presenting its thickened mantle.

Even the seemingly exposed and naked molluscs such as the slugs, squid, cuttlefish and nudibranchs are less of a tender mouthful than you may think. A slug that one moment is sliding along minding its own business can hunch up into a ball shape in seconds. Its tentacles withdraw and it exposes a large hump of its mantle, which is covered in a thicker leathery skin. This posture also has the effect of making the slug harder to pick up or bite into, especially when it starts to produce a thicker slime. Try picking one up, the goo is like glue.

Some species even supplement this with a little nervous rocking dance; what the purpose of this is no-one is quite sure. Nudibranchs go for chemical defences in quite an advanced way; some of the gaudy colourful species have toxins in their skins, stolen from their prey, and in the most advanced cases the actual stinging organs of their prey, the nematocysts from soft corals, hydroids and anemones, are ingested and sequestered in sac-like organs on the body surface. These recycled defence organs are then released in vengeance on any fish or other predator who ignores the bright billboard of warning coloration.

Try catching a cephalopod and you soon realise the fantastic power of jet propulsion as the beast makes a rapid sprint for safety. This is made all the more effective by the other useful trick up their sleeve, a little puff of ink which rapidly expands to give the escaping animal a smokescreen and leaves the naturalist or predator with nothing but a phantom image of the animal once there. Even the smaller ones, such as the little cuttlefish that often turn up in rockpools, are capable of creating such confusion.

Some of the bivalves are also capable of a trick or two. Scallops can skip or fly when they sense approaching danger, particularly from their arch-enemies, starfish. They create this movement by rapidly flexing their valves and squirting a jet of water from their exhalant valve. It is not the most controlled movement but is an effective way to escape their slow-moving nemesis.

Performing a digestive feat the Sea Slug not only avoids being stung, but after consuming the stinging hydroids, somehow steals the armoury of its prey and incorporates it into its own body.

Ink is a brilliant smoke screen for many cephalopods from Octopus to Squid and Cuttlefish. A squirt of the stuff and any attacker is left biting into the ghost of what was there – the mollusc having deployed its jet to make a speedy getaway.

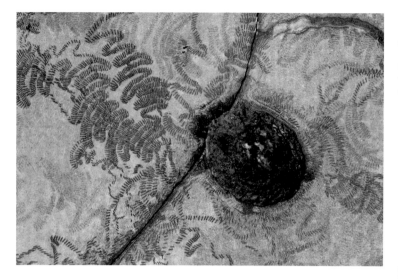

Heading home after being out on the rocks – these hardy snails graze the rock surface of algae. The top picture shows marks made by the scraping radula, the bottom picture shows the trails through the silt, proving these animals get about and are surprising active.

The limpet

You'll find these conical-shelled molluscs on rocky shores. There are several different species, with different-shaped shells reflecting slightly different lifestyles and habitats. Where the rocks are soft, these long-lived molluscs often etch the surface like cookie cutters, and when the rock is submerged and they move about to feed they leave a circular mark – their 'home scar' – to which they return as the tide retreats.

You can test just how loyal they are to these bases by mapping the rock surfaces, and at low tide mark limpets in the study area with a unique coloured mark that coincides with their position. Make careful notes on this. Use coloured correction fluid, oil-based paint or an indelible marker.

When you return after the next high tide, look for your limpets. Have they moved? Have they returned to the same base or moved to another? This sort of study can be carried out over a long period of time and patterns can be observed in animals that otherwise seem stationary and dull to the untrained observer. You can do studies of this nature on pretty much any snail species, terrestrial or aquatic. And if you don't believe that limpets move at all, have a snorkel around your study rock at high tide and you'll see the individuals you are getting to know gliding across the rock surfaces, grazing the layers of microscopic algae.

~ FOOD FILE: THE RADULA ~

A feature unique to molluscs is a file-like, rasping tool called a radula. This structure allows them to scrape algae and other food off rocks, rasp away at vegetation and in some species it is even used to drill through the shells of other mollusc prey. In the cone shells, it is modified into a harpoon to catch fish, and loaded with potent venom that can kill a human!

You'll need a good magnifying lens, and a co-operative animal, to get a glimpse at the hardware. You can watch pond snails as they cling to the underside of the water surface, grazing the film of microscopic plants, or pop one in a jam jar and watch it slide up the side of the glass – if it is feeding you'll see its radula move backwards and forwards almost like a jaw. You should be able to see this even more clearly with a Garden Snail or a slug (the larger the animal the larger its radula) – not only do they have relatively big mouths, but they will eat almost anything.

You will need:

- food processor
- lettuce, grass or cornstarch
- cuttlefish bone
- small piece of Perspex or glass
- paintbrush
- snails

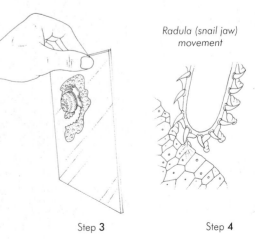

Step **1**

Step **2**

Radula (snail jaw) movement

1. First you must tempt your snail to eat. You need to make a tasty soup – feel free to experiment. I've tried lettuce, and a little added sugar. Place the ingredients in a blender to make a nice runny green liquid. Add a bit of water to thin it down.

2. Paint this thin liquid onto one side of the clear sheet using the paintbrush, and leave it to dry. Repeat this a couple of times to build up a thick layer.

3. The next step is to place your hungry snail onto the glass/Perspex, until it starts to move. Then slowly lift the sheet into a vertical position.

4. Using your magnifying lens, hopefully you'll notice your snail is hunched up and is not moving too fast; instead it will be using its radula like sandpaper to scrape the dried soup from the surface. You can see this action as the snail repeatedly opens its mouth, protrudes its radula and swallows. The snail will also leave a distinctive feeding trail and you'll see the zigzag marking left by the rows of 'teeth' on the radula – this is a pattern that you often see on the glass of greenhouses where the algae has been grazed off by nocturnal slugs and snails.

Step **3**   Step **4**

Deadly drill

You'll sometimes find the shell of a bivalve mollusc such as a mussel on the beach with a neatly bevelled round hole in it. This is the work of the adult necklace shell, a mollusc-eating mollusc. The hole is made in the live bivalve by the necklace shell's radula. It works a bit like a dentist's drill but also dribbles shell-dissolving acids while rasping away to get at the defenceless mollusc inside. The necklace shell's victims are often found together on the beach as the hole alters the fluid dynamic of the shell. Left-hand shells often end up at one end of the beach, right-hand at the other, and predated ones somewhere in the middle.

The neat hole in this clam is a clue to the mollusc's demise – it was eaten alive by a Necklace Shell.

Have heart

If you keep pond snails in a small aquarium or tub under bright light, and make sure they are fed well, they usually grow quickly, but their shells tend to develop a little on the thin side. This is handy for the snail enthusiast, as you can use a magnifying lens and actually see into the shell of a living snail. You'll notice the mantle cavity and, beating away next to it, the two chambers of the heart. Similarly, if you backlight the shell of a banded snail you can see the massive mantle cavity, which acts like a lung.

Meet the bivalves

You will probably be familiar with these animals from the fishmonger's slab if you have not met one yet in person. Scallops, mussels and cockles are all bivalves. The name bivalve refers to the two parts (bi) to their shells, or valves. The largest and most extrovert of this group are the suitcase-sized giant clams, filterers of seawater and gardeners of photosynthetic algae, which they nurture in their mantles. However, the typical bivalve mollusc is a rather unassuming creature. Most live a secretive and enigmatic life, lurking in the oozing mud of estuaries, skulking in the sands and silts, quietly filtering and vacuuming nutrients from water and sediment. But just because you cannot see them doesn't mean they are not up to some interesting things. To get to know them just requires a bit of alternative thinking, a gardener's fork and a soil sieve!

Living filter units: these Zebra Muscles are busy filter-feeding, sucking water in through a siphon.

Secrets of the sands

Forget those irritating little bendy plastic buckets and spades. On your visit to the shore take a sturdy full-sized steel fork and instead of a bucket take a shallow tray or small glass observation tank. Now you're prepared to uncover the odd life forms that hide below the sand's surface. It helps to learn to read the etchings on the sand surface; these often blow the cover of the animals below.

The trick to getting to see the creatures is nothing more than digging gently down with your fork and sifting and sieving through the sand and mud. You can turn up many different kinds of creatures, not just bivalve and gastropod molluscs but crustaceans such as little shrimp-like animals and different types of crabs, echinoderms (urchins and starfish), a plethora of different worms and a fish or two; all of which have rather specialised behaviours and adaptations for the unique challenges this environment offers.

Common are coiled spaghetti-like casts of sand, each paired up with a shallow depression that gives away the presence of Lugworms (*Arenicola marina*). Small star-shaped marks surrounding a single hole reveal where Ragworms (*Nereis diversicolor*) gingerly reached out to scavenge for food (more about these later). Two closely associated holes, one with various patterns of grooves around its mouth, reveal the location of molluscs such as the blushing pink-shelled Thin Tellin (*Angulus tenuis*) which has paired siphon tubes, one long mobile tube that hoovers the surface of nutrient-rich sediment brought in by the tide, and another static one through which water is expelled. Other molluscs leave nothing more than mysterious random hieroglyphics and holes to be pondered over.

The best place to see any activity is down at the water's edge, so the savvy naturalist will follow the retreating tide. As always in the intertidal zone the highest diversity of animals, including those seen least often, are near the low water mark, due to it being uncovered for the shortest periods between the tides. You may observe jets of water being squirted in the air by various molluscs that, when they detect the vibrations of a predator

The best place and time to find sand-living animals like this Rayed Trough Shell is to wait until the lowest tides.

Crabs inhabit many habitats from fresh water rivers and stream to the deepest ocean and every wet place in between. This Ghost Crab is one of the many species worldwide that makes its living on sandy beaches.

(that's you!), will pull their sometimes huge 20–30cm (8–12in) siphons back so rapidly into their shell that any water left in the tubes is forcibly ejected as if from a water pistol, straight up. Depending on depth these can then be carefully dug up and investigated.

The molluscs are mainly bivalves, their shells often modified and streamlined to slip easily through the sand in the activity of burrowing. These animals can be seen almost as living filter units. Some stick their two siphon tubes, inhalant and exhalant, out of their burrow and into the water that passes close to the bottom, where they sit and passively suck this mixture of water and muddy nutrients down for processing, blowing out the waste water. Others are slightly more proactive, with incredibly long and mobile inhalant siphons that systematically work like a vacuum across the surface of the mud. It is these activities that produce the distinctive star-shaped patterns on the surface once the tide has retreated. Gastropods can also be found here; the tiny little spire shells (*Hydrobia* spp.) cruise the surface of mud and sand and leave a detailed weave of lines in their wake.

These squiggles on the sand speak of a Lugworm living beneath.

Etchings of *Hydrobia*

These very important little mud-loving gastropods make up for what they lack in dimensions (approximately 4mm (¹⁄₆in) long) in sheer numbers; they can exist at densities of 60,000 per square metre (foot) and are rich pickings for any bird with the technique to harvest them. The Shelduck, for example, swings its bill from side to side at the surface, sifting hundreds of these little snails up between the tides.

To get a real idea of the numbers, you need to scoop up a handful of mud and water, swish it around and dump the whole lot in a shallow white tray. The quantities of mud in the water should be just enough to leave a fine layer of silt on the bottom when it settles out. Leave this for a few moments. You'll then find that any small animals that were present have started moving. This includes the

miniscule *Hydrobia* snails which, as they start cruising across the muddy tray, leave a trail behind them. You may also stumble across another hidden hero of the estuary mud – a small shrimp-like crustacean called *Corophium volutator*. This little guy lives in shallow burrows in the mud and feeds on detritus washed into and around the burrow entrance. When you are watching wading birds such as yellowlegs, Redshanks or plovers working the mud, pecking frequently at the surface, they will be eating both of these little creatures.

These tiny snails occur in huge numbers in estuaries and salt marshes where they are an important food for many waders and wildfowl.

The Razor Pod Clam – is long, thin and streamlined so it slips easily through the mud and silt when it burrows.

Pretty much any indentation you investigate with your trowel will turn up a bivalve mollusc or two. Some of those with very long siphon tubes can be found at considerable depths and you may have to go down 40cm (16in) or so to get at them, while others will be sitting just below the surface.

One species that will prove difficult to lay your paws on is the Pod Razorshell (*Ensis siliqua*). It and its numerous allied species are nearly all in possession of a huge and very strong muscular foot and an elongated and very smooth pair of shell valves, a familiar presence on many strandlines. These animals are the drag racers of the bivalve world, and when they perceive a predator coming they practically 'sprint' through the sand to get out of the way. The only trick I know that gets them out is the following, but be warned – the success rate of this seems to vary from place to place and season to season. However, when it does work it shows off the impressive abilities of this odd mollusc to the full.

To see this mighty mollusc in the flesh requires a certain amount of cunning. Walking the lower reaches of the beach, look for shallow depressions in the sand or the telltale squirts of water as the streamlined animal pistons into the sand at staggering speed – half its body length a second! Sprinkle salt on the surface above the burrow. This temporarily irritates the razorshell and it will back up. When just enough is protruding make a quick grab at the shell and pull; if you are quick enough you should have it in your hand. If you are too slow the animal will swell the end of its foot in the burrow and will not budge. If this happens give up, and look for another as any further effort could injure the mollusc. Place the razorshell on the sand and watch as it extrudes its long muscular foot. It probes for a while before slipping back under the sand, giving a quick squirt of disapproval before disappearing from sight.

The rare sight of a Razor Clam coming out of its shell – they are often the other way up!

Far from boring piddocks

The main purpose of a mollusc's shell, for most species, is to act as protection and to a degree as support. Some, as we have seen, use highly polished and hydrodynamic shapes to enable them to burrow and move with ease and surprising speed through mud and sand. Those found in more turbulent conditions have developed spikes and keels for friction, helping to keep the animal buried and safe from predators' eyes when the seabed is stirred up a little. But there is a group of molluscs – the piddocks and the shipworms – which have turned their shells into very effective drill bits!

Like other burrow-living molluscs, they hole up and stick their siphons out to filter passing water. What makes them worthy of a separate mention is that their shells, which are reinforced with ribs, spines and teeth, are used to grind away at hard substrates such as rock and wood. The muscular foot provides the twisting motion, with a sucker that protrudes through the permanent hole in between the two valves (this is something you can see if you find washed-up shells of these species; if you hold the two valves together you'll notice they don't appear to fit very well). The shipworm got its name from its habit of boring into the wooden hulls of ocean-going ships and weakening the structure.

The shells of this piddock – a kind of clam, are twisted to form a shape that works like a drill bit, cutting a burrow through rock, wood and hard sediment.

Cockle-flipping

The Common Cockle (*Cerastoderma edule*) is a very common filter-feeding bivalve and can only live within the first few centimetres of the sand. This dumpy mollusc, which is ridged so that it stays put in the sand, doesn't look all that athletic. But place one in shallow water and it will 'leap'! Using the protruding muscular foot, it can suddenly lever against the surface, resulting in a few rolls. Not all that impressive, but of great significance to the cockle as it allows movement to new feeding grounds or even escape from slow-moving predators.

Before you release your cockle there is a little trick you can perform that will give you some idea of how these and their kin actually filter-feed. Pop your mollusc in an observation tank, fill it with clean

seawater and then mix up a milky consistency of brewer's yeast in a separate container. Using a pipette, squirt a little of this above the open siphons and watch what happens! The yeast is sucked in and although water comes out of the other siphon it comes out clean and filtered. This is the reason why some people are nervous about eating bivalves if they don't know where they come from!

Tellin' what it's eating

The Baltic Tellin (*Macoma balthica*) is a very common mollusc on sandy shores (this observation technique can be used for many other related species). Set up a saltwater tank with two-thirds of its depth filled with the sandy substrate from which you extracted your mollusc, then using a piece of clear plastic, you can trap your tellin up against the front glass; this simply stops your mollusc burrowing away from where you can observe it. After a few hours your mollusc will have settled down and burrowed to the depth it likes and hopefully will have extended both its siphons to the surface and will be feeding.

If you add a little dental scale detector (available from your dentist or pharmacy) it will bond to bacteria in and on the sand's surface. In the same way as it bonds to organic material and bacteria in your mouth to show you where you haven't been cleaning, the detector will show you where your tellin has been cleaning! It'll suck up the stained bacteria with its siphon. This is a great way of seeing the animals at work in a way that you really probably wouldn't appreciate otherwise.

Shell-collecting

Many mollusc shells are tough and easily outlast the original occupant. To the naturalist this can be a useful way of getting to know your species, by making a collection of shells. This is commonly associated with a walk along the seashore, but collecting terrestrial molluscs shells is just as rewarding and there are many more different kinds than you may at first think! Storing them is easy and you can mount them in shadow boxes or display cabinets, either gluing or pinning them in place. If stored horizontally, simply subdivide the case into compartments – a good way of displaying many examples and variations of the same species together. As with all collections, detailed labelling of your collection is important – include your name, date and location of your find. Some shells, particularly those with smoother surfaces, show up their colours better if they are wet. This can be recreated permanently by applying a coat of varnish.

The shells of many molluscs, particularly those with long spirals, can be made to reveal their inner beauty by using a file to grind away one side and expose the inner architecture. The grade of file you use depends very much on the thickness of the shell; for thin shells a finer wood file is best, while for

Just a little time on a beach, scurrying up and down the tide line can accomplish a varied collection like this – what better way to appreciate the diversity of mollusc life that exists on almost any shore.

the largest whelks and conches a heavy-duty metal file or even a hacksaw is needed. Use a finer grade to finish the edges nicely and you have a beautiful addition to your collection and a window into the private abode of a mollusc.

Note that in some locations, particularly beaches within national parks and in locations where sensitive protected mollusc species occur, or there is a second-hand market for the shells by hermit crabs, collecting shells may not be a good idea, irresponsible or even illegal.

Drag-hunting

It may seem peaceful and beautiful out there among the rockpools, but animals still die and are killed, and just as there are scavengers on the land, they are in the rockpool too. The most frequently encountered are the crabs, but some members of this clean-up force are molluscs, especially the Dog Whelks. Drag a dead fish or piece of meat along the bottom of an area where the whelks live, then tie down and secure the meat using string and rocks, and watch what happens. Pretty soon every Dog Whelk

Netted Dog Whelks are attracted to smelly baits or carcasses like hyenas around a lion kill.

in sniffing distance – and these animals are very sensitive to tiny concentrations of smells in the water – will move in on the trail, following it to the source. Watch these determined little molluscs as they move with their siphons held high and stiff like masts.

The worm's turn

There is a lot of confusion about worms. We may think we know what they are – those long pinkish things we see when we dig the garden, right? But there are many kinds of worms the world over and, apart from the fact that they are found in damp environments and have a superficially similar elongated body shape, they are very variable. So before we start looking at their behaviour, let's have a quick who's who.

The true segmented worms or annelids (phylum Annelida) include the Common Earthworms (*Lumbricus terrestris*) that live in our compost heaps and herbaceous borders, but there are also hairy ones that look like mammals, there are some with biting jaws that are active predators, some with flamboyant feathery mouthparts, some that come in a bright and vivid selection of psychedelic colours, and others that are architects, building tubular homes for themselves. There are about 15,000 species in the world, including leeches and bristleworms.

Simple but effective – we all owe a lot to the humble earthworm, which helps to make the soil richer.

A true worm shows some sign of segmentation, although some are so bristly that seeing the segments is somewhat tricky. They have well-developed nervous, digestive and circulatory systems and most have bristles along the length of the body called setae, or appendages that can be used for a multitude of functions from traction to breathing.

Flatworms (phylum Platyhelminthes) are not worms at all really. They are flat, have a sac-like gut, and lack bristles. Most familiar are the little black ones a few millimetres long, which cruise around on the bottom of the pond. Some of the marine and tropical rainforest species are bright and bedecked with funky colours; some even look like they are wearing stripy pyjamas. The phylum also includes flukes and tapeworms.

You can catch flatworms by tying a piece of raw liver to some string and dangling it in a pond. Overnight, any flatworms present will congregate and start consuming your bait. You can now transfer them to a Petri dish to observe under a microscope or a hand lens.

Most flatworms are found in water but there are some terrestrial ones such as this hammer headed flatworm from the tropics of Madagascar.

Roundworms or nematodes (Nematoda) comprise 12,000 described species, but this is probably only a very small fraction of the total. They turn up everywhere, but are so small that they mostly go unnoticed. However, spread a little soil or a drop of pond water under a light microscope and I guarantee an introduction to some of these animals.

Nematodes like this Eelworm are some of the most numerous organisms on earth.

The ribbon worms are worth a mention just because they can be found very easily on rocky shores. I remember lifting a rock on the west coast of Scotland to be confronted by a large blob of what looked like strawberry jam! On closer inspection it started to shift; a movement only observable by the glistening of its surface. When I braved prodding it and investigating further, it unravelled to form what could only be described as a slightly animated ribbon of animal. When it was stretched out on the sand, I estimated it at about 5 metres long (16ft long) (and this apparently is a small one)! This was my first experience of a Bootlace Worm (*Lineus longissimus*). Once again, it is only a worm by name, belonging to the phylum Nemertea. Not all ribbon worms are this big. They live in aquatic and mainly marine environments where they are predatory or scavengers, using an eversible proboscis to spear and consume their prey.

This is s small one but bootlace or ribbon worms can achieve huge lengths of over 50 metres (164ft).

The other phyla of worm-like animals are really in the realms of the specialist and are quite hard to recognise. They include the peanut worms (phylum Sipuncula) and the acorn worms (phylum Echiura). Both feed on detritus and live in sediment and under rocks.

Intestines of the soil: earthworms

Earthworms are very, very successful animals, as the fossil record proves. They have been making burrows and holes in planet earth for some 120 million years. These lean, largely unseen, tunnelling machines are incredibly important to the health of your garden and other ecosystems on the planet, and many iconic figures have sung their praises including Cleopatra, Aristotle (from whom I stole the heading above), and Charles Darwin, who spent 39 years studying them. The reason for this high-powered following is that all these people realised earthworms play a very important role in keeping the very soil on which we stand in good condition. When the soil is good so is the plant life and everything that feeds on it.

Part of the reason they are so influential boils down to their numbers. Every acre of grassland can contain up to three million earthworms and each year between them they will turn over and mix up 10 tons of soil! Like little pink rotavators, they aerate and mix up the soil while ingesting dead and decaying matter, breaking it down into a compost-like humus and neutralising the acidity or alkalinity of the soil.

Not the floppy, slimy things you might expect worms have a tougher side to them as well – just run one gently backwards through your fingers.

The ins and outs of a worm

The segments on an earthworm can be seen clearly, dividing the body up. This long, stringy, muscular physique has lots of advantages. Being long and flexible in all directions allows them to squeeze between particles of soil or vegetation. It also allows them to move forward by a peristaltic motion, using a series of muscular bulges. They appear slimy because they secrete a thick mucous from glands on their surface, which allows them to move smoothly through their burrows. At the same time, the mucous acts like cement, binding together loose soil particles, and it slows down the loss of water through their skin when conditions are on the dry side.

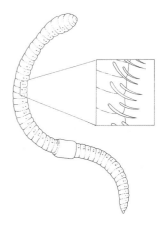

The secret to how a worm gets around can be found in rows along the body. These are the hair-like setae which vary in number depending on the species.

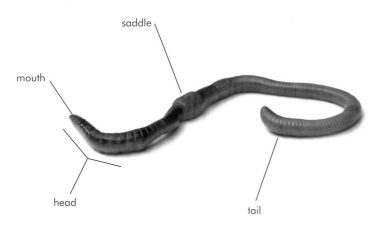

saddle

mouth

head

tail

The **saddle** or **clittelum** is only found in fully grown earthworms and is used in mating and formation of egg cocoons. You can't see them but worms have lots of chemoreceptors – tiny taste buds all over the body but particularly on the 'head end', something like 700 per square millimetre. This is how they find their food. The bristles – setae – are vibration sensors, and help the worm gain traction and grip the walls of its burrow.

Earthworms have no eyes but detect light with light-sensitive cells. Like many nocturnal animals they cannot see red light, so cover torches with red cellophane if you want to watch them after dark. The mouth is the only recognisable feature on an earthworm's 'head' and you have to look very closely to see it, just under the tip of the snout.

Tunnel wizards

Place a worm on the soil and within minutes it has forced its way below, into its dark damp world. Its body design is very flexible and stretchy. The average *Lumbricus terrestris* has about 250 segments, each of which contains lots of ring-shaped muscles that run around the body. There are also long ones running the length of the worm, and it can control these muscles in a very precise way. Its body is also mostly water – 70–95 per cent of your average worm. You can think of each segment as a tough water-filled balloon. The great thing

Hairy worms

Earthworms are lumped together with some 3,000 different kinds of worms known as oligochaetes – the name means 'few bristles'. If you place a worm on a piece of paper you may be surprised at the noise it makes as it moves, not a squishy, slurpy wet noise but a dry scratchy one. If you gently pull a worm backwards between your fingers it will feel slightly less smooth than it looks. Both of these phenomena are caused by the tiny setae (four per segment) that the worm uses to grip the walls of its tunnel to help it move through the soil. It is these same bristles that allow worms to grip the walls of their burrows so effectively, and give even hungry birds a good tug-o-war!

This lapwing struggles to pull an earthworm from its burrow thanks to the worm's many setae.

Earthworms spend most of their lives hidden beneath the surface in their burrows, reaching up when it's dark and the weather's warm and wet to grab leaves and other organic matter to drag beneath the surface and feed on.

about water is that it cannot be squashed. If you squeeze a balloon filled with water, it simply bulges out in another direction; the volume stays the same. Combine the water and the muscles and you have a very impressive digging machine indeed. Using its muscles it can make itself long and thin and squeeze into tight gaps between soil particles. By using muscles and that 'water balloon' effect, it can squash its segments up, causing its body to bulge and push apart the soil. By repeating this action over and over again, the worm can push and pull its way through the earth, making its job easier all the time with its gripping bristles and by producing slime.

Worm-charming

Sometimes you'll see gulls 'puddling' on areas of short grass – sports fields are favourites. The birds appear to be doing a little dance on the spot. The drumming of their feet could sound a bit like rain, and worms tend to surface in the rain, maybe due to favourable damp conditions or to escape waterlogged burrows. Whatever the reason, come to the surface they do, and dinner they become. Bizarrely, humans copy the birds' behaviour in the sport of 'worm-charming'! This involves stamping, patting, prodding and even watering the soil to get as many worms to surface in your square metre as possible. Try it on your lawn or down the park – you may have to wait a while and people may ask questions, but eventually you should be rewarded by a worm poking its head out of the grass!

The private life of *Lumbricus*

Worms are hermaphrodites, each having both male and female reproductive parts. Despite this, one worm does need another. It's thought that they probably rarely bump into each other so having both male and female bits means that each meeting is twice as efficient. If you go out on warm, wet summer nights, treading carefully and armed with a torch covered with red cellophane, search the lawn for worms lying next to each other (these will usually be Common Earthworms, as many other species mate underground).

If you find a pair of worms mating you'll notice they face in opposite directions. They grip each other with the long setae on their bellies, and their clitellums (saddles) both produce a thick sticky sheet of mucous that looks and functions a bit like tape, keeping them close. For the next two or three hours, sperm is transferred, then they go their separate ways. Over the next few days each worm will produce egg cocoons out of mucous from the clitellum.

Just as you peel a jumper off over your head, the worm peels off a band of mucous. Both sperm and eggs are placed in it, and as it rolls over the worm's head the ends seal, to form what for all the world looks like a small lemon. It contains around 20 eggs, of which only a few will hatch.

If you're lucky you might find worms in the throws of mating.

MAKING A WORMERY

Most of what earthworms get up to goes on beneath the surface, out of view. But you can get a window on the worm's world by constructing a wormery.

What you need:
- 2 pieces of clear Perspex 30 × 30cm (12 × 12in)
- 14 small wood screws
- 1 × 116cm (46in) long piece of 2 × 2cm (¾ × ¾in) wood
- 1 × 30cm (12in) strip of wood
- elastic bands
- selection of different-coloured soils (garden soil, sand, potting compost)
- 2 pieces of 30 × 30cm (12 × 12in) black card/paper
- screwdriver
- drill
- scissors
- 5–6 earthworms

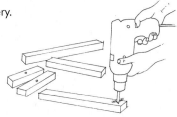

Steps **1** and **2**

1. Cut the long piece of wood to these lengths: 2 × 30cm (12in), 1 × 26cm (10in), 2 × 15cm (6in) and place the pieces together.

2. Using a drill (get an adult to help with this) drill two holes a few centimetres from the corners on each edge and repeat on the other side. Also drill two holes through the shorter 26cm (10in) piece of wood (these will eventually act as drainage holes) and a hole in each of the two remaining 15cm (6in) pieces. These will become the feet of the wormery.

Step **3**

3. Using the 14 screws, screw the whole contraption together as in the illustration. The wood should be sandwiched between the two pieces of Perspex, flush at the edges; the feet should be screwed on last.

4. Add the soils, one at a time in alternating layers, add a few leaves at the top and water lightly.

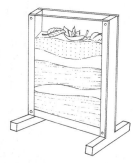

Step **4**

5. Add the worms and place the 30cm (12in) strip of wood on top, using the elastic bands to secure it. This acts as a lid and stops the worms escaping. Then place the two pieces of paper/card over the sides of the wormery and keep in a cool place. Remove the card/paper blinds when you want to observe the worms. Check daily to make sure the soil is damp, but never soggy and wet.

Step **5**

Spineless wonders 🌿 **285**

Roll-ups

Have a close look at the surface of your lawn in autumn and you may notice what looks like hand-rolled cigars stuffed into tiny holes in the lawn. What on earth has been going on here? Look for clues, and you may see that surrounding some of these 'cigars' are worm casts.

Armed with a torch covered with red cellophane (remember worms are blind to red light) go out on a warm, wet autumn night and scour your lawn for worm action. You may find worms mating, and if you are really lucky you may catch a worm reaching out of its burrow with its front end, grabbing a leaf in its mouth and pulling it by its stem into the burrow. It will naturally fold and roll to fit the hole. Now the worm covers it with digestive spittle and leaves it until it has gone nice and soft, before it tucks in.

Worm poo – that's what a worm cast is, soil that's been passed through a worm's simple digestive tract.

Cast away

Those strange wiggly things that spring up all over your lawn are, and there's no polite way of saying this, worm poo! They come from one of two species, *Allolobophora longa* and *A. nocturna*, which between them are responsible for nearly all the worm casts seen in Europe. Other species produce them, but normally in the entrance of their burrows. Worm poo is not unpleasant, it's soil, but with nutrients in a form available for plants to consume – it's become fertiliser.

If you're a proud lawn owner, tolerate the odd worm cast, as if you were to try and exterminate them for their actions you would have to spend lots of money on extra fertiliser, and effort on drainage. Also, no other wildlife would be attracted to your lawn to feed on the worms.

A different can of worms: other annelids

Many other annelid worms can be discovered when you know where to look. In the sea are some free-swimming ones that use their paddle-like parapodia to propel themselves along, using a sinusoidal wiggling motion. Worms that live in tubes, such as the Peacock Worm (*Sabella pavonina*) and Christmas Tree Worm (*Spirobranchus giganteus*), only really reveal their glorious beauty when submerged by the tides. They can be seen if you snorkel, or wade with an underwater viewer.

Worms can be beautiful – these stunning peacock worms are emerging from their tubes in the sea bed to fling their 'arms' around and catch food, looking good while doing it.

These animals are filter-feeders and extend long flowery feeding organs out into the passing current, trapping small animals and other nutrients from the water, and although they look pretty and passive they are really getting down to business. Like microscopic 'pass the parcel', little cilia hairs on the tentacles move particles of food trapped in the sticky mucous down towards the centrally positioned mouth. On disturbance they will surprise with their rapid retreat, like a feather duster disappearing down a drainpipe at speed. It's an essential party trick, as fish like to nip off the tips.

Anyone who still thinks worms are dull would be hard-pressed to continue with this viewpoint having seen some of these tunnel-living marine animals, in particular Christmas Tree Worms which are common sights on tropical and sub-tropical dives, like nervous, gaudy, twin-headed bottle brushes. A misplaced curious finger can ruin the whole psychedelic dream scene, as the feeding heads will be rapidly withdrawn and the top of the tube slammed shut with a little hinged lid. Where once there was a field of bright worms, you are now left with nothing but a regular rock!

Other worms that can be found while exploring the muddy and sandy soft bottom are best seen at low tide, by digging. Look for the burrow entrances and casts of Lugworms – these living vacuum tubes are easily exhumed by digging between the paired squiggly cast (the back entrance) and the conical hole (the front entrance) with a gardener's fork. Pop the worm in a shallow dish of water and you'll see its various features, including the bright blood red pairs of gills that run down its body.

Christmas Tree Worms, each with a double head, are a filter-feeding species of warmer seas. Cast a shadow over them and they vanish into their tubes in the blink of an eye closing a protective door (operculum) behind them.

Worms can be active too, this white ragwort moves across the silt to find a new place to burrow.

Look out for scavenging Ragworms; these pretty animals have all manner of greens, reds and blues in their livery and with their many paddle-shaped parapodia down their sides, resemble old rag rugs. They remind me of Chinese dragon dancers in miniature – 'Dragon Worms' would be just as appropriate a name. Like all soft-bodied invertebrates, they look best in a shallow tray of water. Try to avoid handling them as they have fearsome jaws and will bite!

Other worms you may find on the shore include the noticeable Green Leaf Worm (*Eulalia viridis*). In spring in the UK you may find their egg clusters – more visible than the worms themselves, these are little bright green spheres of jelly, like soft green marbles.

Another active worm, the Green Leaf Worm uses its many 'leaves' to paddle its way across the rock pool.

Right: In Rwanda it is possible to see a Giant Earthworm of this size. Some have been reported to grow to around 6m (20ft).

Getting botanical
Plants and allies

In this chapter we explore the non-animal kingdoms of life. The most important of these is the plants, but we will also take a whistle-stop tour of the mouldering masses known as fungi. To any but the most dedicated botanist, these may seem a little dull compared with the animals we've been looking at in the rest of the book. Certainly it is more of a challenge to get nose to nose with a Polar Bear than it is to shove that same nose into the stamens of a Buttercup; and the heart may beat out a bossa nova when a Tiger turns up, while a Tiger Lily inspires little more than a sigh. Having said that, the biggest living thing on earth is a plant. The Giant Redwood (*Sequoiadendron giganteum*), which grows from seeds like pinheads, can weigh 2,145 tonnes and reach 83m (272ft). In terms of value, the dead giant club mosses have, over millions of years, formed the world's coal seams, making most of human industry possible. These are the life forms on which all the other, more energetic lives on earth depend. How can they be dull? Let's see what happens when we look at them a little more closely and apply that greatest of the naturalist's skills, a different perspective.

Right: More than just old dead leaves: leaf litter is where many ecological stories start.

Left: Giant Redwood trees – the biggest living things on earth.

Why care about plants?

At one level, the question of what plant grows where is down to soil type and prevailing weather conditions, but if you look at the bigger picture, plants shape an ecological community; quite simply, they determine the presence of other life.

I first realised this when I was eight years old and crazy about insects. Plants hadn't happened to me yet, I hadn't met one that really impressed me and my eyes hadn't opened to a world that moved at a barely perceivable pace and at a microscopic level. As far as I was concerned, plants were the sorts of things Mum and Dad messed around with – fuchsias and petunias in hanging baskets – and not cool at all.

But the green things were waiting for me just round the corner, and funnily enough, that first awakening came not from the sort of triffid-like plant you might expect to capture a young lad's imagination, but from the very same fuchsias my mother was lovingly cultivating. They were being nibbled by something large and caterpillar-like, and late one night while scrutinising the herbaceous border with a torch I found one of those dream bugs – a finger-sized monster Elephant Hawkmoth caterpillar. It was a beauty, and I collected another and another until I had several jam jars full of fuchsia leaves and big caterpillars. The catch came when Mum showed her displeasure at my caterpillar farming, because I was forever raiding her plants for fresh leaves to feed my captives. The only solution was to cultivate a few fuchsias of my own. I soon became quite a fuchsia expert and enjoyed watching my plants grow almost as much as I enjoyed watching the caterpillars eat them.

As my career as an entomologist continued to blossom, I realised that an understanding of the foodplants on which various butterflies and moths depended at different stages of their lives was critical to an understanding of the animals themselves – not only how the plants affected the distribution of a species, but also how the intricacies of the plant's life cycles fitted into the lives of the animals. Now that really pushed my buttons. Much later, when I worked in butterfly houses and farms, and also as a field

Rafflesia, or Stinking Corpse Lily, produces the world's largest known flower, measuring about 1m (3ft) across. It is a parasitic plant and has no visible leaves, root or stem. When in bloom, it exudes a horrible smell of rotting meat, which attracts insects to pollinate it.

biologist studying the ecological requirements of threatened fritillary butterflies, the majority of my work became botanical. It doesn't end there, either – as a naturalist you soon find out that some birds roost only in certain trees, some bats feed on specific fruits and flowers, some beetles eat the bark of only one species of tree. And so it goes on.

I guess what I'm trying to say is, whether you like it or not, plants will affect your life, especially if you plan to spend any time in the great outdoors. In South America, knowing the subtle distinguishing features of regular grass species and Razor Grass (*Scleria scindens*) can mean the difference between a comfortable day in the field and looking as if you have been in a fight with a Jaguar. The same goes for an unfortunate incident ,when one of my friend's colleagues mistakenly used the leaf of a potently poisonous plant as toilet paper – you can guess the rest.

I got to know plants because it helped me find and look after caterpillars. Many species of caterpillar only eat the leaves of one or two species of plant; they have quite specific dietary needs.

One of the world's most venomous plants

In Australia there is an infamous plant called the Stinging Tree (*Dendrocnide moroides*) which leaves all those who come into contact with it in no doubt as to how it got its name; for several days, every time you bathe or otherwise get the irritation wet, you are reminded of the full intensity of the original contact. I have even heard of a guy dying having swum in a jungle pool full of dropped Stinging Tree leaves! So it makes sense to learn something about them.

The Stinging Tree in rainforest, Tolga, Atherton Highlands, Queensland, Australia.

Plants... what are they?

Plants can be divided into two main groups: the non-vascular plants and the vascular. The major difference between them is vascular plants have an internal circulatory system while non-vascular plants don't.

The first and more primitive are the non-vascular plants, which have no special water-carrying vessels within their tissues; they include 14,000 species of moss, 9,000 species of liverwort and 25,000 species of algae. These plants are examples of some of the earliest to have existed. Plants evolved in water and many of the non-vascular plants, particularly the algae, still grow there; others, including many mosses and liverworts, have moved on to land but retain a connection with moisture and cannot live in dry places.

The second group, the vascular plants, have true leaves and roots and an internal circulatory system of vessels which transport food, water and other necessary nutrients and chemicals around the plant's tissues.

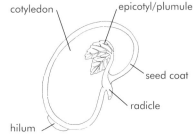

You can see cotyledons 'in action' if you split a bean into its two white, fleshy halves; between them is a white miniature plant ready for the off. Look closely, and you will see all the wonderful details of the embryo plant – the leaf structure and lots of tiny perfect veins.

Vascular subdivisions

The vascular plants can be further divided into the ferns, of which there are about 10,000 species, and the most recent and successful of the plant groups, the flowering plants, whose 250,000 species can be further subdivided into the monocotyledons and the dicotyledons. Monocots include grasses, palms, lilies, irises and orchids and are thought to be more recent in evolution than the dicots, which are more numerous and include all the other flowering plants. The name comes from the food-storage organs in seeds, called cotyledons, and all flowering plants have either one (mono-) or two (di-). These will shrink as the embryo grows, as they power its shoots and roots both upwards and downwards. The size of the cotyledons varies from plant to plant and gives some idea of a seed's needs – the smaller it is, the nearer the surface it needs to be planted. If a seed is positioned too deeply in the soil its stores will not have enough energy to push the shoot to the light.

Military Orchid – a monocotyledon.

Rhododendron – a dicotyledon.

Unassuming little plants

mosses and liverworts

These low-growing plants produce a microclimate all of their own. They exist in sponge-like mats that hang on to moisture and tend to become shelter for many smaller moisture-loving animals. Mosses have little hollow leaves and big cells which act as moisture-storage capsules – a desirable property for animals that need moisture. In the wild you can thrust your hand deep into a sphagnum clump, often up to your elbow, and pull up a plant to see that only the very tip is actually alive, green and growing – but the rest of the plant, long since left behind down below, keeps on soaking up moisture and providing for the living bit upstairs.

Some mosses have strange, uprising structures that look almost flowery. These contain the plant's sexual organs and require moisture to enable the male cells to swim toward the female organs. Once the female parts are fertilised, they produce thin filaments with a bobble on top – this is the sporing body and in some species, when it dries out enough, it becomes explosive, blowing up and scattering its contents to the wind. These spores can be cultivated in the same way as ferns and if you scatter them on to damp blotting paper, you can watch them germinating under a microscope.

Liverworts have a similar life cycle but keep an even lower profile than mosses. Look at their leaves and you will see they have cup-and-envelope-like structures from which free-floating buds called gemmae are produced. These often rely on water splashing or flowing over them, and from each one a little liverwort will grow.

These are the fruiting bodies of a moss. The spore capsule at the end will lose its cap and open up to release the tiny spores when it dries out and conditions are right for the spores to blow away.

 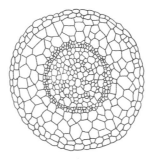

Opened moss capsule and cross-section of the moss stem.

Simple plants
ferns and their fronds

Look under the fronds of mature fern leaves and you may well see lots of odd little 'C'-shaped brown things. These are the plant's spore capsules (spores are a fern's or fungus's take on seeds) and if you shake the frond when conditions are right, the spores will show themselves as a light dust. To understand the fern's life cycle, harvest some of these spores by shaking them into a Petri dish of sterile, moist soil. Cover them and keep them in a light place out of direct sunlight. Eventually the spores will produce a strange, single leaf-like disc called a prothallus, which will quickly wither away. Don't lose heart, though, as something else has been happening at a microscopic level. On the surface of the prothallus, egg and sperm cells will have developed and, in damp conditions, the sperm will swim to the eggs and fertilise them. From these female organs a small fern will grow. The sperm's dependence on a film of water to make its way to the egg explains why ferns cannot live far from moisture.

If you start getting a feel for ferns, mosses and other moisture-loving plants, you can always keep some of your babies and grow them on, recreating the high-humidity environment they need in a bottle garden. Both the club mosses and the horsetails also have a prothallus stage in their life cycle.

Fern – young fronds unfolding. Their reproductive parts develop on the leaves.

The spore-bearing 'cone' stems of the Giant Horsetail).

Things to do with spores

Ferns need to release their spores only when the environmental conditions are dry and right for 'wafting'. If you look at a sporing body under a powerful hand lens, you will see that along the edges are thick cells which act as water-sensitive triggers. Shine a torch or some other weak heat source on the spore capsule and you should be able to persuade the fern to snap open and fling out some spores. Drop some water on the plant and the capsules shut up tight again.

Horsetails also distribute themselves by using spores, but their sporing bodies are tall, free-standing, cone-like structures. If you shake out some of the spores and look at them under a microscope, you will see an odd structure with four straps attached to it. These straps act as springs and generate movement, which helps with dispersal. Heat them up and the straps coil up quickly; moisten them, and they uncoil again.

The big green life-making machine

photosynthesis

Lie on your back in a deciduous woodland in the springtime and you will feel it. As you gaze up through the back lit tessellations of leaves with the sun dancing over their perfect surfaces, you are looking at a multitude of invisible light-processing machines, an almost magical alchemy that harnesses the sun's rays and gives planet earth the green mantle that supports everything else upon it. Without it we wouldn't have wildlife to look at – let's face it, we wouldn't even be here to contemplate the fact!

The way we see green plants gives us our first clue to what's important to the life-giving process. We see green because it is the wavelength of the spectrum that the plants use least – it is surplus to their requirements, so is reflected back at us. The reds and blues are snatched away from the rainbow and used to drive the process that turns sunlight, carbon dioxide and water into the fundamental foods that kick off pretty much every food chain on earth. That process is called photosynthesis, which means 'putting together with light'.

When the sun's rays hit our planet, 99 per cent of them are absorbed by the land or the sea or even reflected back into space, but the remaining 1 per cent is trapped by little green blobs, processing units called chloroplasts that are held within a plant's tissues. Within these is contained, neatly packaged, a chemical called chlorophyll and this wonderful stuff is the key to life on earth. The magic that now takes place can be represented by this formula:

CO_2 (carbon dioxide) + H_2O (water) →
O_2 (oxygen) + $C_6H_{12}O_6$ (sugar)

What happens in the chloroplast is that various pigments such as chlorophyll B, carotene and xanthophylls absorb the sunlight and channel it to the other form of chlorophyll, chlorophyll A. Here it releases charged particles called electrons, which are then replaced by the hydrogen atoms in water. The outcome

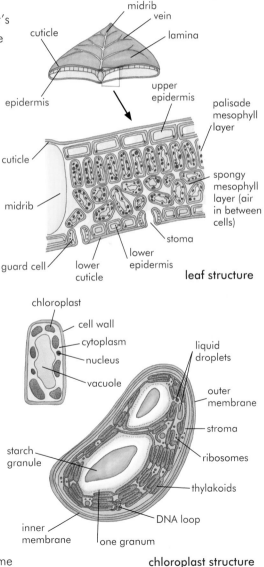

leaf structure

chloroplast structure

of this part of the process – called the 'light reaction' – is that water is split into its component parts, hydrogen and oxygen. The oxygen is released as a gas, while the hydrogen goes on to meet carbon dioxide in the next phase, called 'carbon dioxide fixing'. Eventually this produces simple sugars such as glucose and fructose ($C_6H_{12}O_6$), any excess of which may be stored as starch, or the more complex sugar sucrose ($C_{12}H_{22}O_{11}$). These energy-rich chemicals are stored in the tissues until they are needed, a bit like a battery pack that the plant uses up when it has to grow or when it is too dark for photosynthesis to take place. The gases involved in photosynthesis enter and leave the plant via little pores called stomata in the leaves and green stems.

While plants are busy building sugar and giving off gases useful to the rest of life on earth, that very same life on earth is selfishly taking those sugars and reversing the process by means of something called cellular respiration. This means extracting energy from the sugars by breaking them down, using oxygen and giving off water and CO_2 in the process – exactly the opposite of what happens during photosynthesis. This chemical reaction, which occurs within every green leaf in the world, is the reason why nothing can live without plants; the greenies are inextricably linked to the lives of every breathing thing by that rather dull-looking bunch of numbers and letters printed on the previous page. It's also the main reason why environmentalists get angry when people start cutting down the biggest concentrations of photosynthesises on the planet; the rainforests.

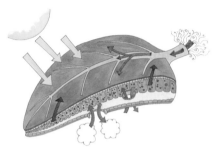

This diagram shows photosynthesis and transpiration in action. The leaf takes in sun and water and produces sugar and oxygen.

The pallid flesh colour of plants such as broomrape stand out due to the lack of chlorophyll. This is because they are parasites, stealing their living from the roots of others.

Partial photosynthesis

Greater Dodder stems parasitic on nettles, Slovenia.

Not all plants use photosynthesis to produce all their food. Some, like the mistletoes, are semi-parasitic and although they do photosynthesise, they also feed off their host plant – often an oak tree in the case of mistletoe. Others have no chlorophyll at all and are totally dependent on their hosts. Examples of this include the toothworts (*Lathraea* spp.), broomrapes (*Orobanche* spp.) and dodders (*Cucusta* spp.), most of which have an odd pinkish hue to their stems and only really exist above ground as flower spikes.

~ THE BUBBLES OF LIFE ~

All around us plants are producing oxygen as a by-product of photosynthesis – which is handy for the rest of us, as we need oxygen in order to breathe and carry out all of life's processes. We simply cannot get away from the fact that plants are the foundation stone of life on earth.

Most of the time you obviously can't see a plant produce a clear, colourless gas and release it back into a cocktail of more clear, colourless gases. The exception is when the plant lives underwater, and the following experiment is a great way to witness the production of one of the most useful end products of photosynthesis.

Step 1

You will need:

- a clear jar
- a funnel
- a test tube
- a handful of pond weed
- a long match or wooden splint

1 Fill the kitchen sink with water. Fill the jar with water, add the pond weed and place the funnel upside down over the jar.

2 Immerse the test tube and turn it upside down without breaking the surface of the water, making sure there is no air trapped inside it. Place the test tube over the submerged neck of the funnel.

Step 2

3 Lift the ensemble out of the sink and place it on a light, sunny windowsill. As the pond weed begins to photosynthesise, you will see small bubbles forming on the surface of the leaves and rising in streams to be collected in the inverted test tube. When the tube is half full (the time this takes depends on how bright the light is and how warm the water) light the match, quickly blow it out again and, while it is still glowing, lift the tube and thrust the splint into it. The splint should burst back into flame. This is proof that the gas your humble pond weed has produced is life-giving oxygen itself!

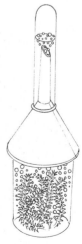

Step 3

~ CATCHING THE LIGHT ~

'Potato' is just the name we have given to the underground reserve or tuber of a certain kind of plant, and the tuber is the place where this particular plant stashes away vast quantities of energy-rich starch. If you cut into a potato and drop some iodine solution on the exposed tissue, you will notice that it quickly turns from yellow to a blue/black colour. That is the test for the presence of starch. Once you are convinced about that, it is time to try an experiment with the living light-processing units on a handy houseplant.

You will need:

- a houseplant you don't mind damaging
- a few small sheets of black plastic (tearing up a large bin bag will do nicely) or kitchen foil
- a few clothes pegs
- 2 Petri dishes or saucers
- small quantities of methylated spirit and iodine solution
- a heatproof beaker
- a small saucepan
- a pair of tweezers

Step 1

1 Wrap some but not all of your plant's leaves in the plastic or foil, holding this in position with clothes pegs so that no light can penetrate to the surface of the leaves. Leave on the windowsill for two days.

2 Take one leaf that has been starved of light and one that has been left as normal. Put some methylated spirit in the beaker, stand this in a pan of water and heat until the spirit boils. Remove from the heat and, using tweezers, dip each leaf into the hot water for a minute. Then dip them into the spirit and leave until they are almost white.

Step 2

3 Put each leaf in a separate Petri dish or saucer, add some iodine and watch what happens. What you should observe is that only the leaf that was exposed to light turns blue/black, showing that starch, the product of photosynthesis, is present. Only when light could reach the plant's chloroplasts did photosynthesis take place.

Step 3

The way to grow

tropism

Plants have never been known for their speed of movement. Although there are exceptions, for the most part they live at a pace at which change is observable only if we film them and speed up the footage. When this is done, they take on whole different personas. A questing bramble stem becomes a serpentine arm reaching out and dragging itself forward on its ratchets of thorns, an unfurling branch of leaves becomes a flapping ballet of salad and a blooming flower looks like a pyrotechnic display.

Growth occurs at specific points within the structure of each plant and different species with different ecologies grow in different ways. To demonstrate this, take a dish of germinating grass seed and one of germinating cress seed. Once they have grown up a few centimetres (an inch or two), take a pair of scissors, give both a buzz cut and wait to see how the seedlings respond. You will find that the grass picks up where it left off, just as it does when you mow your lawn, but the poor cress will never recover from your barbarity. Why? Well, it's all about the position of growth areas known as meristems. In the grass, the meristem sits in a band of tissue between the roots and the shoot, so when the shoot is trimmed or grazed, the growth cells remain intact. Grasses evolved in ecosystems where they were continually eaten by grazing animals, so this is a vital survival strategy; it also explains why some members of this huge family provide us with such sumptuous, springy lawns. With the cress, the growth zone is right at the very tip of the shoot; remove this and the shoot is doomed.

We all know how fast grass grows: when shoots are grazed or trimmed growth cells remain intact, so it regrows. It grows from special growing points at the base.

The fastest growers

The fastest-growing plants in the world are the bamboos, king-size grasses that are capable of reaching for the light at an incredible 1m (over 3ft) a day. But even this is not quite fast enough to register with the human eye. Instead we have to rely on indirect evidence of plant growth.

The fastest-growing plant in the world – bamboo – you can almost see it grow.

Growing plants move towards and away from stimuli, following the lead given by the part containing the meristem. Directional growth occurs when cells grow and multiply faster on one side of a shoot than on the other, in response to increased levels of growth hormones. This phenomenon is called tropism, from the Greek meaning a turn, and there are many different kinds. One of the commonest is phototropism, which means it is connected with light. This may be either a positive move, as seen when germinating seedlings grow towards the light, or a negative one, as in the seed heads of Ivy-leaved Toadflax (*Cymbalaria muralis*), which 'self-plants' its own seeds in dark crevices and so grows away from light.

These seedlings are leaning to the right – the predominant direction of the light source – a classic example of phototropism.

Responding to other stimuli – nastic movements

Movements that are not directional and are not controlled by growth are known as nastic. Examples are the everyday opening and shutting of flowers such as daisies and celandines in response to light intensity; and the similar response of crocuses to temperature change. Some of the quickest plant movements are those observed in so-called 'sensitive' plants such as *Mimosa* species. On physical contact they will close up their leaves as a way of avoiding looking succulent and being grazed by passing herbivores. Another interesting nastic movement is that exhibited by the carnivorous plant the Venus Flytrap (*Dionaea muscipula*), whose scientific name means 'mousetrap of Venus'. While they don't actually stretch to consuming mice, they have, like nearly all carnivorous plants, evolved to live in boggy, nutrient-poor habitats and supplement their diet with the bodies of the animals they catch.

Some plants such as the Venus Flytrap have adapted to nutrient-poor soils by obtaining their protein from the tissues of animals, caught by their ingenious devices.

Venus Flytraps have a pair of specialised leaves that form the famous trap. The surfaces of these two leaves are held open, exposing a red surface and producing sweet nectar, both of which are attractive to insects. On each of these leaves are three hairs, and when an insect lands on them, the pressure on the hairs causes the leaves to snap together.

If you hold a bright light behind a Venus Flytrap in a dark room and look along the back

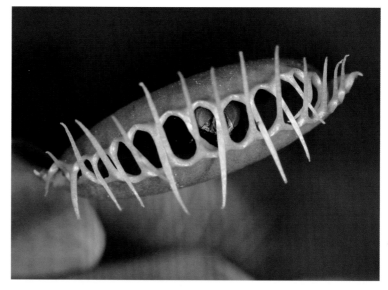

lit 'jaws', you will see the three trigger hairs on each half. If you touch one of them with a small paintbrush, nothing happens, but touch two at the same time and bang, the trap shuts tight. If you touch the same hair twice within a certain time frame – say 20 seconds – the leaves will also snap closed. This complex mechanism helps the plants to save their energy, as it has to be a fairly big insect to trigger the hairs in such a way. Such an insect's struggles will further stimulate the trap to stay tightly closed. But if the plant is fooled by a small ant or even by a curious naturalist poking it with a paintbrush, the trap will shut only halfway. This allows the insect to escape, as the plant has sensed that the returns it will provide are not worthy of the effort involved in digesting it.

Even in the heat of noon, the Sundew's glandular sticky hairs continue to twinkle with dew – hence its name – luring insects to their death.

There are many other sorts of carnivorous plants, including pitcher plants (*Napenthes* and *Sarracenia* spp.), sundews (*Drosera* spp.), butterworts (*Pinguicula* spp.) and bladderworts (*Utricularia* spp.). If these fascinate you the way they fascinate me, you are in good company – they were also a great favourite of Charles Darwin. You can experiment to see what stimulates the leaves to curl and release the digestive enzymes by placing various foods on them. Try leaves, grass, sweets, insects and small pieces of beef. It is the protein itself that gets the juices flowing.

Empty pitcher

If you come across pitcher plants in the wild (not in UK), try emptying their contents out into a tray. I am often surprised by the prey items and other residents I find. As well as the partially digested remains of many types of insects, I have decanted the bodies of frogs, lizards and even a small mouse. The real mind-boggler is that, in some plants, despite what must be strong digestive juices, I have found the living larvae of mosquitoes and other water-dwelling insects. Spiders and tree frogs also sometimes live on the doorstep of death by hiding out in the humid confines.

A Shrub Frog sits inside a Pitcher Plant, Malaysia. Frogs are one of many small species that can be found inside these flowers.

~ PHOTOTROPISM ~

You can demonstrate positive phototropism by growing seedlings on a windowsill: the shoots will visibly grow towards the light; turn them around and they grow back the other way. Put little foil hats on some and they will continue to grow upright in response to gravity (see Geotropism), oblivious to the light stimulus the others are following. You can take this to extremes by challenging a germinating seedling to a bit of an obstacle course.

You will need:
- a cardboard box with a lid, and extra cardboard to make baffles
- matt black paint
- a planted bean or sunflower seed

1 Insert a couple of baffles into your box and paint all the inside surfaces with matt black paint to absorb any light that might otherwise be reflected.

2 Make a small hole at one end of the box with the end of a sharp pencil. Set the box on its side with the hole at the top. In the bottom place your bean or sunflower in a pot of damp soil. Now put the lid on the box.

3 Wait and watch what happens as your seed germinates and the shoot, complete with its light-sensitive cells, goes searching for the pinhole that is letting in the light it needs.

You may wonder how a plant can make all this growth and all these contortions without any light, but remember the seed itself is an energy-storage unit designed for just this sort of exertion. The energy was put there by the parent plant.

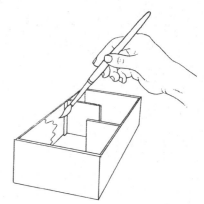

Step **1**

Step **2**

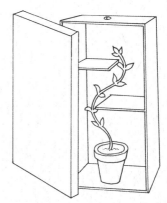

Step **3**

~ GEOTROPISM ~

This is a pattern of growth influenced by gravity. Plants need to know where to put their roots, and the pre-programmed response that allows them to ground themselves firmly in the earth is sensitivity to gravity.

You will need:

- a jam jar with a lid
- a wire support
- a bean
- glue
- cotton wool

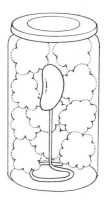

Step **1**

1 Glue the wire to the inside base of the jar and put the bean on it. Pack some moist cotton wool into the jar and replace the lid.

2 Wait for the bean to start throwing out a shoot and a root. Make a note of the direction of the root's growth.

3 Then place the jar in the dark and turn it upside down. Watch what happens to the root. Repeat and repeat, and note that whichever way the jar is turned, even in the absence of light, the root will always grow downwards.

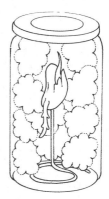

Step **2**

Step **3**

Propagules, batteries and time travel
the secrets of seeds

Seeds really are extraordinary things: they are compact, easily stored and capable of surviving drought, sub-zero temperatures, even fire; in their dried state they resist attack from mould or bacteria and, despite being energy-packed food sources, they manage to avoid the gaze of many seed-eating animals by strategies such as camouflage. They are full of surprises, too: some are dynamic – they can fling, ping, pop and whizz – while others are incredibly good at doing nothing at all. The record for any living thing sitting around doing nothing is claimed by the seed of an Arctic Lupin (*Lupinus arcticus*), which was successfully germinated after more than 10,000 years!

Seeds come in a variety of sizes from the coconut-like product of the Giant Fan Palm (*Lodoicea maldivica*), which weighs 20kg (44lb), to the minute seeds of some species of tropical orchid, with 992.25 million per gram (that's more than 28,000 million per ounce)! But whatever their size, their mission is simple. They have to go forth and find the best place to grow, away from competition, predators and detrimental climatic conditions. They can do this in a huge variety of exciting ways, and they do it all while they are technically dead, as their cells show no signs of metabolic activity.

Equipped with everything it needs to start a new life, a seed simply sits and waits for the three triggers it requires for germination: the right temperature, enough water and the right situation such as the correct soil. Until these things come together, the embryo is held in a state of suspended animation by its dehydrated state – typically less than 2 per cent of a seed's weight is water, compared with 95 per cent in a thriving herbaceous plant. When all the conditions are right, the seed coat lets the water in, at a rate which varies from species to species. In some species the coat also needs to be digested by bacteria in the soil first, or the seed needs to be subjected to an abrasive action called scarification, caused by passing through the gut of a bird or mammal. This is one reason why some seeds sit inside tasty fruits that advertise their presence to animals in the hope of being eaten, dispersed and deposited in a ready-to-grow condition in their own little pile of manure.

The seed of a sycamore tree has a wing that allows it to spin in the wind and travel far from its parent. Here, moisture has triggered germination, and the root, powered by the seed's energy store, heads for the dark.

~ WATCHING GERMINATION ~

We generally get to see only about half of the process of germination, because the other half goes on below the surface of the soil. This is where basic hydroponics comes in handy. Hydroponics is the technical term for growing plants in a liquid medium, and it provides the interested botanist with a window on this stage of a plant's development.

You will need:

- a transparent glass vase or jar
- a sheet of paper
- cotton wool
- a seed (ideally a large one such as a bean, maize or sunflower)
- chopsticks or a similar long implement (optional)
- graph paper (optional)

1 Place the paper in the vase or jar, fitting it closely to one side. Stuff the rest of the vessel with cotton wool.

2 Push a seed into the gap between the paper and the glass, using the chopsticks if it helps. The seed should stay towards the top of the jar so that you can see a decent amount of root growth; if not, stuff in some more cotton wool.

3 Add enough water to reach the seed but not cover and drown it.

4 If you want to record the daily growth of the roots and the shoot, stick a strip of graph paper to the outside of the vase and mark it each day with a line and a date.

Steps **1–4**

Agents of dispersal

To prove how good seeds are at their job description, look at your dog's coat or your own socks next time you have been out for a walk through long vegetation. Many plants were using what we proudly call space-age technology before humans had worked out how to carve a wooden club. Velcro was not a product of space exploration, it was invented by plants – just think how those burrs and grass seeds have attached themselves to anything that is fluffy or woolly. They use a multitude of little grappling hooks that vary in design and number but work on similar principles.

If you're still not impressed, scrape the dirt off the bottom of your shoes and mix it up with water. Then prepare a sterile seed tray with soil that you have microwaved to kill off any other seeds that might be there. Pour your shoe mix onto the tray and place the whole lot in a clear plastic bag. Leave it on a windowsill and watch what happens.

Cleavers, or Goose Grass uses the Velcro technique.

Dandelion parachute hairs open out in dry weather for wind lift-off.

Opposite: Willow herb attaches plumes of hairs to its seeds so they blow away in the wind.

Suck it up: transpiration

Even though vascular plants have a circulatory system that superficially resembles ours, they have no pumping equivalent of the human heart to force their life blood through their veins. Instead they rely on transpiration. This entails losing water, usually through the stomata, and although this may sound like a bad thing, it is actually a very useful 'suction pump' process that keeps the plant alive.

A lone water molecule is a rare thing in nature – their natural affinity means that one always attracts another. So as water molecules evaporate from the surface of a leaf, other molecules move from neighbouring cells to take their place; then others move in from the next cells along, and so on and so forth until you get to the source of the water – a feat of micro-plumbing called the xylem vessels, which link the leaves to the roots, where the water enters the plant, and in which water exists as a very long chain of molecules. As these are pulled out of the top, they drag others in the chain up behind them, and with the water move the minerals, vitamins and all manner of other substances necessary to the plant's health and survival.

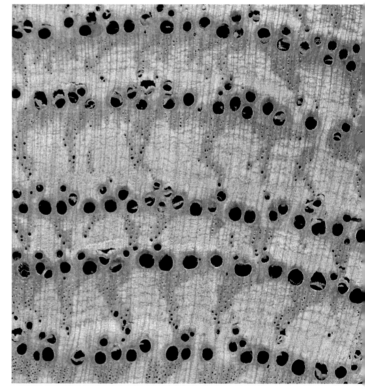

This cross-section of a branch of English Oak shows quite clearly the micro-plumbing of the xylem vessels.

In addition to trying the experiment opposite, you can measure the rate of transpiration by putting the stem of a cut plant in a jar containing a measured amount of water and adding a thin layer of cooking oil, which forms a seal and prevents any water being lost by evaporation. Water will therefore be lost only through the stomata, so any decrease in the water level will be down to transpiration alone. By timing the experiment and marking the water line on the side of the jar, you can measure not only the volume but also the rate of water lost.

If, when the experiment is finished, you put the leaves on a piece of graph paper and draw around them, you can even work out the rate of water lost per square centimetre or square inch of leaf. Warming up the room or standing the experimental set-up in front of a fan will enable you to assess how temperature and air movement (wind) affect the rate of transpiration.

~ AN EXPERIMENT WITH TRANSPIRATION ~

Although transpiration is invisible because it operates at a molecular level, its influences are often felt – it is the end result of transpiration that forms clouds and leads to daily deluges of rain above the tropical forests, for instance. An average 15m (50ft) broad-leaved tree can transpire up to 260l (58gal) of water an hour! On a more accessible level, the effects of the same process can be witnessed on your own windowsill. Try this:

You will need:

- 2 glasses
- a few drops of food dye
- a long-stemmed, pale-coloured flower such as a carnation
- a sharp knife or scissors
- sticky tape

1 Fill the glasses with water and put the dye in one of them.

2 Using the knife or scissors, split the stem of your flower up the middle to about the halfway point. Wind a little tape around the top of the split to hold it together.

3 Place each half of the split stem in a glass of water, move the whole set-up on to a windowsill and leave for a few hours. (The sunnier and hotter it is, the faster the water will evaporate from the petals and leaves and give you a result.)

4 When you return, you should see that the half of the flower whose stem is in the coloured water has turned the same colour as the dye, demonstrating that the water that has been sucked up as the plant loses water from its petals is directly connected to the half of the bottom of the stem sitting in the coloured water.

5 For an even simpler demonstration, put a humble white daisy in a pot of watered-down blue ink for a few hours and watch it turn blue.

Step **2**

Step **4**

Sun catchers

leaves

It's funny how we all notice leaves as soon as they drop off the tree and start clogging up drains and blowing around the garden. But how good are you at telling different types of tree apart by looking at their leaves? Why not start a collection? Press leaves between the pages of heavy books (see p.328), make prints by pressing them into the surface of modelling clay, or turn them into lacy skeletons. Impressions of leaves can be made in a variety of ways: if you simply brush a leaf with shoe polish, making sure you work it well into all the leaf surface, then place it on a piece of paper oily side down, cover with another sheet of paper and rub firmly all over the surface with your finger or a sponge, you should get a good imprint. This works best with older leaves that have fallen from the tree, as they tend to have more prominent vein structures.

Another excellent way of achieving an image for your records is to make a leaf skeleton (see p.314) and then hijack a photographer's dark room. Place the skeleton on a piece of photographic paper below a table lamp, turn the lamp on for five seconds, remove the leaf and place the paper in developer for three minutes, soak in a water bath for an hour and hang up to dry. You should now have a stunning photographic image of your leaf's veins on a black background.

A leaf print shows off to perfection a leaf's 'micro-plumbing' – its characteristic vein pattern and shape.

Autumn colour

It's not called 'the fall' for nothing. The most obvious feature of this season is that the leaves of deciduous trees drop off. Millions upon millions of leaves that have served their purpose in the spring and summer, attracting the sunlight and housing the chlorophyll that are essential to photosynthesis, must now be dispensed with. During a harsh winter, when it is difficult for a tree's roots to obtain enough moisture from the cold soil, it can't afford to lose too much water through transpiration. So the leaves have to go.

They don't go quietly, though. Their parting gesture is the most flamboyant display of their lives. The key factor in producing the reds, golds, oranges and even purples that people cross continents to see is decreasing day length. This triggers the plant cells to produce a hormone called abscisic acid, which in turn forms a corky layer of cells (the abscission layer) at the base of the leaf, where it joins the twig. This acts as a seal across the vascular tissue that normally carries water and nutrients into the leaf and takes waste products away. As a result, the leaf effectively starves, the stem that holds it to the plant weakens and it is eventually carried away by a gust of wind.

This may sound like an unhappy ending, but before the final curtain, the leaf makes a fantastic exit. Because it isn't getting the nutrients it requires, it stops manufacturing chlorophyll. In fact, chlorophyll is an unstable substance that breaks down and needs to be replaced all year round, but cooling autumn temperatures speed up the decomposition, replacement chlorophyll isn't produced and the green colour that has covered the landscape for the last six months fades away.

Pigments on display, revealed by the death of chlorophyll.

In its place appear more stable pigments – the carotenes, which can be further divided into the xanthophylls and the carotenoids. These are responsible, respectively, for the yellows and the reds and oranges which have been waiting in the wings while chlorophyll has been stealing the show, and which now take centre stage. Another group of chemicals, the xanthocyanins, which produce the rich crimsons and purples, are created by the last late-seasonal gasps of photosynthesis, but trapped in the leaf by the abscission layer as it firms up.

Fabulous foliage

The maples (*Acer* spp.) that cover New England are spectacular examples of autumn colour, but British natives such as dogwood (*Cornus* spp.) and Wild Cherry (*Prunus avium*) may steal a last curtain call. The production of xanthocyanins is encouraged by dry sunny days and cool, crisp nights, so we enjoy the most spectacular autumn colour under these weather conditions.

~ PLANT PLUMBING ~

These lacy skeletons are a great way of showing off how the leaf works. The skeleton consists of the 'plumbing' used to supply the leaf with water and to transport the sugars made in the leaf cells during the summer to the rest of the tree.

You will need:

- leaves
- washing soda at the concentration of approximately 40g per litre (6oz per gallon)
- a pair of tweezers
- a paintbrush

1 Heat the soda mixture until it nearly boils, take it off the heat and soak your leaves in it for 30 minutes.

2 Using tweezers, carefully remove the leaves and gently wash them under a tap. The soft parts will fall away (this can be further encouraged by teasing them with a paintbrush).

Steps **1–2**

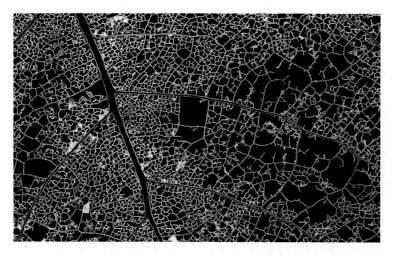

This close-up photo shows just how fragile the skeleton of a hybrid Black Poplar leaf is; it is not dissimilar to a map or plan of a city.

314 🌿 *The complete naturalist*

The rude bits

flowers

We give them to our loved ones, wallow in their scent and select them for our gardens because of their colour and aesthetic appeal, but flowers do not exist for our benefit. They are the sexual organs of flowering plants, and the ones we notice most are those used as billboards to catch the attention of passing insects, birds and to a lesser extent mammals such as bats, who have a very serious job to do. The plants use all sorts of lures such as gaudy colours, flamboyant petals and enticing perfume, and as if that were not enough some species offer a little gift in return, in the form of energy-rich sugary nectar.

This is all about the flying animals collecting the flowers' sexual packages of pollen and transferring them to another plant so that it is fertilised. Some flowers – certain species of orchid, for example – are just a big con act; they look like a sexy example of a female insect, tricking the poor males into trying to mate with them. Before the males give up frustrated, they will have done just what the flower wants, in return for zilch!

From the naturalist's point of view, flowers are also useful ways of identifying the plant that is producing them. When studying them, look at the structure of the flower, and at the size, colour and number of the petals and internal organs. Bear in mind that even though many flowers contain both male and female parts, some flowers are only male, others only female.

Busy bees: this bumblebee is working hard collecting pollen and nectar from this flower.

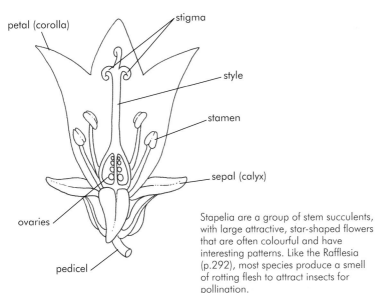

petal (corolla)
stigma
style
stamen
sepal (calyx)
ovaries
pedicel

Stapelia are a group of stem succulents, with large attractive, star-shaped flowers that are often colourful and have interesting patterns. Like the Rafflesia (p.292), most species produce a smell of rotting flesh to attract insects for pollination.

We tend to overlook some other forms of flower, such as those of certain species of tree and grass, because they don't need to be showy. They are simply launching pads for the pollen, which is cast out to the mercy of the winds in the hope that it will find another plant to fertilise. The odds in this game of botanical roulette are skewed back in the plants' favour by the production of millions upon millions of tiny, windborne pollen grains, the stuff of hay-fever sufferers' nightmares!

So how do you tell an insect-pollinated flower from a wind-pollinated one? Well, the insect-pollinated flowers are likely to be large, showy and brightly coloured. The anthers (the part of the stamen – the plant's male reproductive organ – in which the pollen ripens) are often hidden within the flower in a way that facilitates contact with the insect or other pollinating agent. These flowers generally produce fewer pollen grains, whose structure tends to be spiky and complex, so that they will stick to the pollinator and be carried away to be deposited on another flower. The stigmas (female sexual organs) of these flowers, which need to receive the incoming pollen, also tend to be strategically positioned inside the flower, to ensure maximum contact with the pollen-dusted pollinating agent.

Wind-pollinated flowers are often inconspicuous, with green petals and no obvious scent. The anthers are large and only loosely attached to the flower, so that the slightest breeze shakes the pollen free. The flowers often dangle, with the sexual parts exposed; the stigmas are feathery to maximise the surface area that will catch pollen. These plants produce large quantities of smooth, light, easily windborne pollen grains.

The flowers of Common Bent Grass release a lot of light, small pollen grains that are easily carried away on the wind. Wind-pollinators need to produce a huge amount of pollen to ensure the success of at least some pollen grains reaching the stigmas on nearby flowers. The diagram below shows a cross-section of a grass flower.

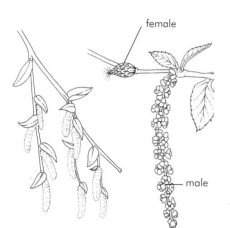

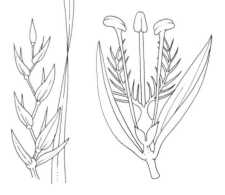

Some trees produce male and female flowers. The male flowers, or catkins, all hang in a group, like lambs' tails. One catkin can release millions of pollen grains which are transported by the wind. Some will, with luck, land on the stigmas of a female flower of another plant.

The oldies

trees

Trees are arguably the oldest living things on the planet, the record being held by a species of bristlecone pine (*Pinus longaeva*) in California whose age is confirmed at over five thousand years! The high altitude and cool conditions they grow under contribute to this longevity, but even in less favoured parts, trees are some of the oldest living features of nearly all landscapes, and there are a few tricks that a naturalist can use to calculate their age.

The best known way of ageing a tree is called dendrochronology and unfortunately involves taking a saw to it and cutting it down! Looking at a cross-section of a tree which has been smoothly cut can reveal a number of interesting growth features. The patterns of rings and lines seen are the direct result of how the plant has grown. Just beneath the bark is a layer called the cambium, the cells of which retain the ability to keep dividing in three different directions: inwards to form xylem vessels that become the 'wood' cells, outwards to form the phloem cells essential for food transportation, and sideways to increase the circumference of the tree. (If you think about it you will realise that trees have to grow out as well as up, otherwise they would eventually become top heavy, unstable and fall over.)

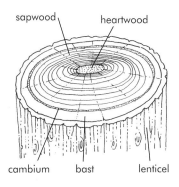

Each year during the growth period of the tree, the cambium cells multiply and add a new layer. It is these layers that produce the concentric circles or rings. Count these and you will have a pretty good idea of how old the tree was before it was cut down. Other features of interest are the medullary rays – spars of tissue that connect the inner and outer layers of the trunk, enabling waste material from the outer layers to be dumped in the centre dead wood and adding further support to the plant in much the same way as the iron ties used by builders shore up and add strength to walls.

As the tree grows outwards, the cells in the centre die off and become impregnated with a hard, tough substance called lignin. These dead cells are far from useless to the tree, for they are what gives the trunk its strength. The very centre of a tree trunk is known as the heartwood. Occasionally a weakness in the tree, such as an injury or a lost limb, allows fungi and other decomposing organisms to gain entry to the dead tissues. This can have mixed results, for although eating away at the heart is obviously bad news for the tree, it also forms cavities which are then colonised by bats or nesting birds and are in themselves ecologically very important.

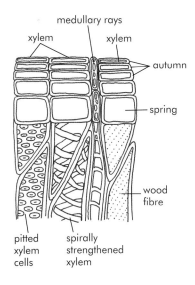

~ ESTIMATING A TREE'S HEIGHT ~

There are a number of ways of estimating the height of a tree that are easier and quicker than shinning up it with a tape measure. The first relies on comparing the tree to something whose height you know – in this case, a friend.

The friend, stick and pencil method

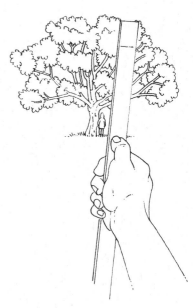

1 Measure your friend's height, then ask him/her to stand at the base of the tree in question.

2 Hold a measuring stick out at arm's length and line the top of it up with your friend's head. Mark a line on the stick that coincides with the position of your friend's feet at the base of the tree.

3 Without moving, raise the stick so that the top lines up with the top of the tree, remembering to keep your arm locked out in front of you. Make a mark on the stick that relates to the position of the bottom of the tree.

4 These two marks give you a comparison between the height of the tree and the height of your friend. Dividing the larger figure by the smaller shows you how many times higher than your companion the tree is. So if the tree is 18 times the height of your friend and he/she is around 1.75m (5ft 9in) tall, the height of the tree is approximately 31.5m (103ft).

The belly method

This only really works with smallish trees. You need an obliging friend and a stick that is about 2m (6ft 6in) tall.

1 From the base of the tree pace out 27 paces of equal length, then push your stick in the ground. Take another three paces and mark this spot.

2 Lie down at this spot with your eye as close to the ground as you can get it and ask your friend to move his/her finger up and down the stick until the finger lines up with the top of the tree from where you are lying. Mark this position on the stick.

3 Because the tree is 10 times further away from you than the stick, multiply the distance from the bottom of the stick to the fingermark by 10, and you have your estimate of height.

The pencil method

Probably the easiest of all. You need a pencil, a tape measure and a co-operative person.

1 Hold the pencil up at arm's length and walk to a position where the bottom and the top of the tree correspond to the top and bottom of the pencil.

2 Rotate the pencil through 90° so that it is parallel with the ground, keeping the bottom of the pencil lined up with the base of the tree.

3 Get your assistant to walk away from the tree at right angles to your position, stopping when he/she reaches a point that lines up with the end of your pencil. The distance between your assistant and the base of the tree will equal the height of the tree.

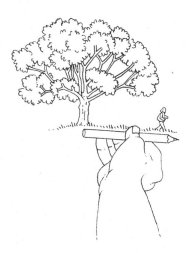

The tangents and protractors method

This is the most complicated but also the most accurate, as long as your tree is growing on level ground.

1 Make a simple clinometer by attaching a length of wood to a protractor, suspending a plumb line from the centre of the wood so it dangles across the protractor, then sticking a nail or screw into the wood at either end of the protractor to act as sights, as on a rifle.

2 If you want to keep the maths simple, walk away from the tree until, by lining up the two sights with the top of the tree, you get a plumb line reading of 45°. The distance from you to the tree plus your own height above the ground equals the height of the tree.

3 If you have a head for numbers, you can use the following formula to work out the height of the tree from any distance. If a is the distance from observer to tree and b is the angle shown on the clinometer, then the height of the tree = a x the tangent of b. The tangent of the angle b can be found from basic mathematical tables. The reason we use 45° to keep it simple is that the tangent of 45° is 1. But sometimes other vegetation may obstruct the view of your tree top at the desired distance, so being able to do the maths gives you more flexibility!

Measuring tree canopies

Mapping the shape of tree canopies is an excellent way to add more information to a tree study – it helps to assess how much sunlight is reaching the forest floor, the density of tree growth in relation to other species, and so on. Mark the position of the trunk of the tree on a piece of paper and then pace out the distance to the edge of the canopy in different directions. Note this down on your plan. Repeat this as many times as you like to get an accurate to-scale map.

Another way of estimating a tree's age that doesn't involve taking a chain saw to it is based on an observation by tree scientists that, no matter what the species, all trees in a temperate locality grow at around the same rate – approximately 2.5cm (1in) in circumference every year. So measuring the circumference of the tree in centimetres and dividing by 2.5 will give you an estimate of its age in years. (If you are measuring in inches, then the number of inches = the number of years.)

Tree skin: learning about bark

Just as birds have distinctive plumage and butterflies have species-specific wing markings, so trees have their own unique profiles, leaf shapes and bark textures. Some have such a distinctive general impression, size and shape that, as with the 'jizz' of birds, you can identify them by profile alone. However, no one tree is exactly like another and unless it is on its own in a sheltered spot, it is unlikely just to be growing straight up. Even among members of the same species, growth and overall shape are affected to varying extents by environmental factors, such as prevailing wind direction, unique to each location, and by competition with neighbouring trees.

Leaves are one of the most useful ways of identifying trees, but at certain times of the year, they may be missing, and although there will often be a few withered 'hangers on' or even a litter of leaves, seeds or fruits at its base, this isn't always the case. Similarly, the profile of a tree may not be obvious – when you are close to it in woodland,

The shape of this wind-blown tree indicates the direction of the prevailing wind, which has altered the way the tree has grown over time.

for example. So you may have to fall back on the texture, colour and patterning of the bark to make an accurate identification.

Tree bark performs a protective function, bearing the brunt of any attack, whether from insect, mammal, bird, fungus or plant – it is a suit of armour, a flak jacket and a bumper all in one. In some species it also takes on specialised tasks. Some of the eucalyptus trees in Australia and Indonesia, for example, are pyrotechnicians designed to survive and benefit from forest fires. They produce light, highly flammable, resin-rich bark that burns quickly when a fire sweeps through. This saves the trees from the potentially fatal effects of slow, hot-burning fires and also serves as a trigger for their fire-stimulated seeds, which start germinating when all the competition has been burnt away.

Other species – particularly in the tropics, where there is a huge number of lichens, algae and other potentially harmful epiphytes and parasites, but also familiar temperate trees such as beeches and birches – have developed very smooth bark or bark that is loose and peels off continuously, renewing the surface and making it hard for any other species to get a root hold on their surface.

On top of these protective uses, in young or green-stemmed trees, bark may be another organ of photosynthesis – some species have air holes called lenticels in the bark that work in the same way as the stomata in the leaves. All these roles, in conjunction with characteristic growth features such as fissures, cracks, branch and leaf scars, combine to form the patterns, textures and colours that make each species unique.

Even if a tree is dead, removing the bark will probably tear the roof off the homes of many smaller creatures that live underneath it. It is also bulky and messy to store and transport.

A much better way to learn about bark is to start a collection of rubbings or moulds and use them as the basis of a journal or scrapbook of species characteristics, such as profiles, height, leaf shape, seed and any other information you can glean. Learn as many of the identification features as you can and test yourself when you are out walking to keep your skills honed.

The bark of this Cork Oak is such marvellous armour that we have used it for centuries to keep bacteria and the like out of wine.

Mangrove trees that grow in seawater have to protect themselves from salt damage. Some concentrate the salt in the bark or in leaves that are about to peel or drop or just excrete it through their leaves. Those with special breathing roots that grow up out of the mud have corky, water-resistant bark to protect them from chemicals in the salty mud.

~ RUBBING IT UP THE RIGHT WAY ~

Bark rubbing was the activity that started me tree spotting with any accuracy as a kid, although as I write, I am sitting in a garden in Miami and I have no idea what any of the trees around me are – I may have to start on these now. You see these skills never become stagnant, you can continue applying them throughout life!

If you have ever done brass rubbing, then you will be familiar with the simple principle involved in bark rubbing. Traditionally your drawing material would be a ball of a black waxy substance known as a cobbler's heel ball (for blacking the edges of shoe heels), but a wax crayon of a colour that contrasts with your paper is a perfectly good substitute. I generally use greaseproof or regular typing paper, at least A4 size (that's about 11½ × 8in) – this allows you to collect a representative section of the bark's pattern – but you may wish to experiment.

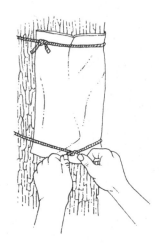

Step **1**

You will need:
- a piece of thin paper
- strong tape or pins to fix the paper to the tree
- a wax crayon

1 Attach the paper to an interesting piece of bark, ideally one that has no lichens or other growths. Attaching it firmly is critical, as you do not want the paper to move once you have started rubbing.

2 Using the side rather than the tip of your crayon and applying gentle but constant pressure, rub over the paper, always moving in the same direction if possible. The crayon will pick out the higher relief of the bark underneath, and you will get a pattern that represents its textures.

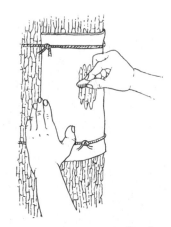

Step **2**

3 Make a note of the species name, location and the height at which you took the rubbing (in some trees the texture of the bark varies with height and age).

4 You can also make rubbings of leaves and add them to your tree log (no pun intended). You can even frame the best ones as natural artworks.

Step **3**

~ MAKING AN IMPRESSION ~

This requires a little more effort than bark rubbing, but the results are spectacular and well worth it in my opinion.

You will need:

- a large lump of modelling clay
- a robust box
- plaster of Paris

1 Knead the clay so that it is malleable, soft and free of air bubbles. This makes it much easier to work with and gives you a better impression.

2 Firmly press the clay into the bark. Try to maintain a thickness of at least 1–1.5cm (about ½in) of clay and don't let the edges taper off. Your mould will be fragile if the clay is too thin.

Step **2**

3 Peel away the clay, which will have the texture of the bark impressed upon it. Place it gently in the box so that you can carry it home undamaged.

4 Back at base, place the mould on a work surface with the textured side uppermost. Using more modelling clay, make a dam around it at least 1.5–2cm (1½–2in) higher than the mould. You can choose at this point whether to make a curved cast, like the profile of the tree, or a flat section, but a curved one always looks more impressive.

5 Mix up some plaster of Paris (see p.85) and pour it into the mould. Leave to set for a few hours and then carefully lift and peel away the modelling clay to reveal your bark cast, ready for display. If you want to bring out the textures even more, paint the cast with a water-based coloured wash. Use bright colours or try to recreate the natural ones, it's up to you.

Step **4**

6 If you are going to make a lot of casts for a collection, it is a good idea to decide on standard dimensions, as it makes them easier to store and/or mount. And why stop at casting bark? You can try making impressions of many things, even leaves and seeds.

Step **5**

~ STUDYING PLANTS IN SITU ~

Earlier on I mentioned some work I once did on rare butterflies that involved studying the vegetation in their habitats. The following are just a few of the common ways of recording and standardising our observations – it isn't good enough to walk about and say, well it looks like there are a few violets here and none over there, for example. You need to have some numbers to back up your statements.

The quadrat

This is a simple device that divides the habitat up into manageable chunks – you simply count the number of species represented within the frame and extrapolate over the habitat as a whole. To get a fair representation of a habitat, you should obviously collect as many sample quadrats as possible. For greater accuracy, this needs to be done randomly – which really means throwing your quadrat over your shoulder so that you have no influence on the process and don't bias your data!

You will need:

- 4 pieces of wood 30cm (12in) long
- 4 pieces of string
- 8 drawing pins
- nails or screws

1 Nail or screw the wood together to form a crude square frame.

2 Mark the 10 and 20cm (4 and 8in) points on each side and pin the string tightly across the frame from each point, dividing your big square into nine smaller, equal squares.

3 This will make a perfectly good quadrat, but if you want something more robust and longer lasting, replace the string with wire and attach it by drilling fixture holes through the wooden frame. Instead of nailing the wood together at the corners, use screws and wing nuts – this allows the frame to be collapsed for easy transportation in your field bag.

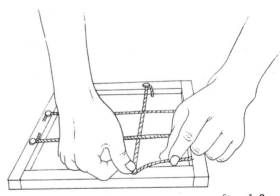

Steps 1–2

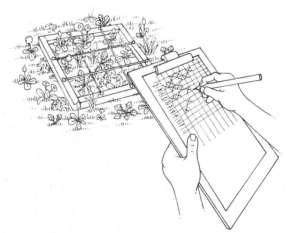

Step 3

Getting botanical 🌿 **325**

You scratch my back

surprising plant relationships

Although the co-evolution of flowers and insects millions of years ago is the plant/animal relationship that most regularly touches our lives – and without it we wouldn't have fruit, seeds or honey – there are many other examples of animal and plant lives that have become closely intertwined.

Ants and plants have symbiotic relationships that can be very complicated and beautiful. These plants are known as myrmecophytes or 'ant plants' and they belong to a variety of plant families. One ant plant is the Bull's Horn Acacia (*Acacia sphaerocephala*), a thorny bush that lives on the African plains, which come complete with a huge diversity of bush-eating browsing mammals such as antelopes. This sounds like a bad thing for the bushes, but the acacia has come up with its own security force in the shape of some aggressive little ants that live within it. This is no casual arrangement, but a formal binding contract that benefits both parties in many ways. The acacia's thorns have swollen, hollow bases in which ant-colony life occurs. Food is laid on, too, in the form of a super-sweet sap that oozes out on demand from extra-floral nectaries on the plant's stems, and protein in the shape of little bag-like pouches called mullerian bodies on the tips of the leaves.

In return for this feast, the ants defend the plant. Just try touching, even breathing on it, and out they swarm from holes in the thorns, armed and ready to defend the colony. Believe me, any naturalist who has accidentally bumped into one of these bushes regrets it – they hurt! The ants also do the gardening, cleaning the leaves of any fungus or harmful growths, and they patrol the ground looking for germinating seeds, which may be competition for the parent plant; these are nipped in the bud before they can get off the ground.

Other myrmecophytes include the hydnophytum plants, which have swollen stems that look like footballs. The ant species that live here collect lots of debris as they forage. This is discarded in the chambers, and the plants grow internal roots that tap into this source of compost. The same principle goes for the *Dischidia* species, which have hollow, bean-like leaves instead.

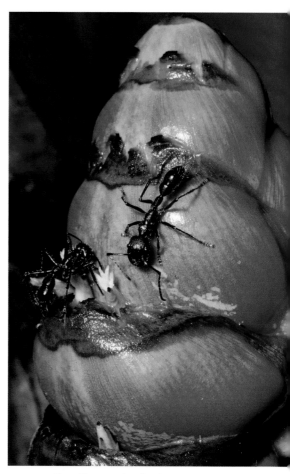

Giant Ponerine Ants feeding on *Heliconia imbricata* plant in a tropical rainforest.

Studying plants at home

Before we get carried away with the idea of studying and preserving plant specimens, it is important to remember that plants are just as vulnerable to over-collecting and damage as animals. Many are protected by law. I realise that many lifelong passions for plants have been born from collecting a fistful of flowers for Mum, but the responsible naturalist should always collect as few reference specimens as possible. Digital photography is an excellent alternative that will enable you to identify many species without even touching the plant.

Having just sung the praises of the technological revolution, I have to say I'm a stickler for the old methods. Getting your hands dirty, touching, smelling and even being stung makes each experience unique and more likely to stand out in your memory, which serves you a lot better than simply pressing a button. On a more practical level, if the plants you are looking at can only be identified by counting the flower bracts, measuring the stamens or observing subtle details such as how hairy the undersides of its leaves are, you really need a specimen. Look at the diversity of grass species in the average meadow and try working them out with nothing but a **camera**. Traditional skills will also stand you in good stead as a back-up when it is bucketing down with rain and you cannot use your camera, the batteries are dead or you haven't brought the camera with you.

At any time in the field you may stumble across a plant that you have never seen before or that looks a little different. You don't have your field guides with you, so what do you do? You could pick a leaf or flower and pop it into a pocket, but it may get damaged and, especially in hot weather, will soon wilt and wither to a pathetic, floppy, unrecognisable mass that is no good for anything other than compost. The secret to getting the most out of your plant specimens lies in keeping them fresh for as long as possible. That is why every botanist worth his or her salt always carries a **plastic bag**. It not only protects the specimen, it also slows down water loss (which is what makes the plant floppy in the first place – see Transpiration, p.310) and you can write notes on the bag itself with an indelible pen for later reference.

Traditionally, field biologists would go one step further by carrying a slightly bulkier and less crushable item of equipment known as a **vasculum**. This is a rigid box lined with layers of blotting paper or plastic between which you sandwich your various samples. You can buy them from scientific suppliers, or make your own from a plastic freezer bag or the sort of box that Chinese take-aways come in.

This is a simple experiment that can be done at home, comparing cress seedlings grown in normal light, left, with those germinated without light, right. The paler colour of the seedlings on the right indicates a nutrient deficiency, or lack of chlorophyll, from no sunlight. If you place the plant in sunlight after, it will recover.

Tips on preparation

It's important to keep specimens fresh until the last minute before you work with them or preserve them. On returning home from an expedition, the busy naturalist may have a number of urgent matters to attend to, so the less demanding plants tucked away in their plastic bags and vasculums may be overlooked. The trick is to buy yourself time. The moment you get home, take the plants out of their containers, rinse them in water and pop them in a vase until you are ready to look at them properly. Transpiration continues after the stem has been cut, so water is drawn up the vessels, leaving nothing but air behind. Even if with every good intention you thrust the stems into water, a bubble of air will eventually block the way, forming an air lock and killing the plant's tissues. If, on the other hand, you remove a section from the bottom of the stem, you hopefully chop away that upwardly mobile air bubble and transpiration can continue unhindered. In other words, your plants stay fresh!

If you are going to dry or press your plants (see below), ensure that they are clean and free from debris and insects (a large, well-camouflaged fleshy caterpillar can really mess up a pressing and the process isn't very pleasant for the insect either). Remove as much moisture as possible by dabbing and blotting the specimen.

Sometimes it is necessary to put the water back into a plant in order to observe a structure or even to take a section to view under a microscope. This can be done with varying degrees of success by placing it in hot water. Leaving a soft, delicate succulent in warm water for a few minutes will probably do the trick, while tougher characters may require a gentle simmer for half an hour or so. The emphasis is on simmer, as a vigorous boil will turn your specimen into soup!

As soon as you get home rinse your specimens, trim them and pop them in a vase, or dry them out and press them. Then you can identify them using a field guide.

The big squeeze – pressing plants

The traditional way to preserve botanical specimens is in a flower press. Despite the name, these are useful for squishing and preserving all but the most robust parts of a plant. They come in different sizes and are basically pieces of thin but stout wooden board with blotting paper between to absorb moisture. It is this that dries the plants out and is the principle behind their preservation. The press is clamped shut by numerous threaded screws and wing nuts or straps.

You can purchase the real deal from any biological or scientific supplier, but sandwiching your plants between sheets of blotting paper and wedging these between the pages of heavy books works well. If you are using books, do try to remember which ones you put your plants in. I am still coming across plant pressings that have been waiting to be discovered for 20 years or so betwixt the pages of *Encyclopedia Britannica*!

Pressing plants does have a couple of limitations. The specimens can lose their original form and become distorted; they will also fade to a greater or lesser extent, depending on what colour they were to start with – green is the most unstable and therefore the worst offender – and how fast they are dried out in the press. Drying is best done quickly, so speed things up by keeping the room as dry as possible and changing the blotting paper every couple of days, to ferry away the moisture. The fleshier the specimen, the more water it will contain, so the more frequently you should change the paper. As a rough guide, most plants will be dry in two to four weeks.

It may seem a little extreme, but that handy kitchen tool the microwave oven is also useful for drying plants. Make a small press without any metal parts and pop this in the oven for a couple of minutes on a medium setting, being careful not to let everything overheat. Remove the blotting paper, check how it's going and repeat with fresh paper until you have a satisfactory specimen.

You can mount dry and flat plant specimens in numerous ways, but you should always add data such as date and location. Giving the limitations mentioned above, photos, drawings and notes on the colours of the living plant are also very useful additions to your record.

If you don't have an actual flower press, you can always press flowers in a notebook. Place the flowers or leaves in the notebook, add additional weight on top, anything heavy – big books are good, and your specimens will dry in 2–4 weeks.

A flotation tank

By using the flotation method you can really get to see, appreciate and learn about the wonderful structures of the numerous species of seaweed.

Specimens that have become floppy and soft, or delicate aquatic plants such as seaweeds and pond weeds can be preserved too, but getting their position correct without the support of the water they grow in can be tricky. So the clever naturalist uses just that – water. Plop your plant into a tray of water and slide a sheet of stout card or paper into the water under the plant. Slowly lift the card until the plant is lying in a layer of shallow water, then use a paintbrush or needle to slide it around until you are satisfied with the arrangement. Carefully lift the card out of the water. Your specimen should now be in its final resting place. Allow it to dry for an hour or so and then gently place it in a press with a layer of absorbent paper on top. Keep pressure to a minimum, and the specimen should not shift its position.

Fungi

Wander through a deciduous wood in the autumn and you will be trudging through a carpet of leaf litter, the home of one of the most extraordinary groups of organisms in the living world. Turn over a handful of leaf litter and you will find fluffy but frail strands of lace; look at the bark of trees and you may see tough corky layers of bracket fungus; and of course glance in dark damp corners and you will find them alive with the fruiting bodies that we call mushrooms and toadstools. These are all forms of fungus, and I call them organisms because they are so weird that experts can't even agree as to whether or not they are plants – but generally it is accepted they belong to their own unique kingdom.

Fungi can be split into two classes – the familiar kind that I have just described are the basidiomycetes, while the others are the 'sac-like' fungi or ascomycetes. These include the yeasts that make so much of baking and brewing possible, and those that feed on human skin and cause such afflictions as athlete's foot. One thing they all have in common is an inability to photosynthesise and produce their own food. So all fungi feed on either living or dead organic material. If it is dead they are described as saprophytes; if it is living, they are parasites.

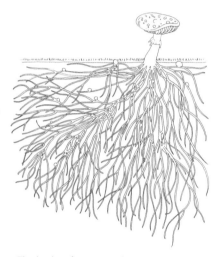

The hyphae form an underground network so vast that there are several tonnes of them per hectare or acre.

Whether they are categorised as plants or not, fungi do not have cells like a plant. Their smallest units are elongated, enzyme-oozing tentacles called hyphae, which infiltrate the food substance and form a net-like arrangement called a mycelium. Hyphae vary in thickness, with the smallest less than a millimetre (¹⁄₂₅in) in diameter. But what they lack in stature, they make up for in numbers – there may be 70,000m of them in a single gram of soil (about 1,200 miles in an ounce), amounting to several tonnes per hectare or acre. The hyphae's task is to move through the cells of their host plant, breaking down tough material, collecting nutrients and turning the one-time stars of the ecosystem into nothing more than fungus fodder. If they weren't around to digest the vast quantity of detritus that Mother Nature produces, the place would be a mess.

Like many fungi, the Fly Agaric has a relationship with trees, in this case birch (*Befula* spp.), depositing certain nutrients into the soil in exchange for carbon from the tree's roots. Find a birch and you might just find the Fly Agaric.

To cap it all: fungus reproduction

What we call mushrooms and toadstools may be just the visible fruiting bodies of the fungus, but what a variety there is – something like 6,000 species in the UK alone. You simply cannot get bored with fungi. Don't get sidetracked by worrying whether or not you can eat them. Just revel in the eerie black bootlaces of Honey Fungus (*Armillaria mellea*), the massive Giant Puffball (*Calvatia gigantea*), which can grow to 60cm (2ft) across, the suggestive form of the Stinkhorn (*Phallus impudicus*), which one early mycologist found so offensive that he coyly drew it upside down. And remember that what you are seeing is only the reproductive part of a massive subterranean being. Under that dead tree stump there may be 15 hectares (37 acres) of mycelia, which may weigh 100 tonnes and be 1,500 years old!

The role of these fruiting bodies is to reproduce by means of scattering tiny, lightweight capsules of genetic material called spores. Autumn is the high season for this, because fungi, both above and below the ground, need lots of moisture in order to grow. For most of the year they lurk out of sight in whatever wet areas they can find, but when water becomes abundant, they are out there ready to breed.

The most familiar way of releasing spores is via a mushroom-like arrangement. The characteristic 'cap' is the fungus's way of pushing its body up out of the dark into the air, giving its spores the best chance of catching the wind and being dispersed far and wide. The umbrella shape is no accident, protecting this living dust from the elements. Puffballs are an exception to this rule – they expose

Velvet Shank Fungus, De Heen, Netherlands. This large collection of fruiting bodies of this velvet fungus means that there is a huge network of hyphae within the tree helping itself to nutrients.

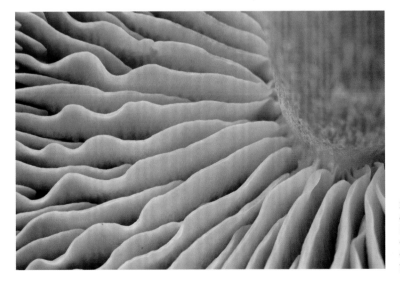

Sculptural beauty: get in close and look at the sporing bodies and you'll be in a better place to understand the ecology of a mushroom. The distinctive shapes, colours and spore prints will also help you identify it.

High spore count

Part of the secret of fungal success lies in their particularly efficient way of distributing themselves around the place. All species, from the tiny moulds to the Giant Puffball, produce spores that, being so small, get everywhere. Your average mushroom can produce a staggering 30 million spores (these are what makes up the 'dust' that is visible when you make a spore print, see p.335), while a 30cm (1ft) Giant Puffball really does throw its numbers to the wind, producing a mind-blowing 7,000,000,000,000 spores! Fortunately, only a minute percentage of these wins the lottery of life, otherwise we would all be living on Planet Puffball.

Not smoke but countless spores released from this puffball.

themselves to heavy raindrops and wind, as these are what makes a ripe sporing body 'puff'. That cloud of smoke so enjoyed by many children is actually a puff of spores. Many species of lichen and bird's nest fungus (*Cyathus* and *Crucibulum* spp.) use similarly explosive means of distributing their spores skywards.

Other fungi utilise insects, particularly those that are attracted to the smell of decay. Cage fungus (*Ileodictyon* spp.) and stinkhorns (*Phallus* spp.) generate a disgusting, sickly sweet stench for the benefit of carrion-feeding flies that walk all over the moist spores, transferring them on to their bodies and from there to everything else with which they make contact. What they lack in olfactory appeal to us they make up for in their incredibly weird forms, personal architecture and structure.

To demonstrate how good these spores are at doing what they do, leave a piece of bread or fruit outside. Within days it will be showing a bloom of furry fungus that will eventually produce little black dots of sporing bodies. These are the 'pins' that give these lower forms of fungi the name pin moulds. What they are doing is simply colonising a resource and recycling it, breaking the food down into its component parts. It was a similar incidence of spores getting into places where they weren't supposed to be that led to the greatest accident in medical history. Alexander Fleming was growing bacteria in a Petri dish when he noticed that a fungus had contaminated one of them and killed all the bacteria. That fungus, penicillium, was later to become the source of one of the greatest antibiotics, penicillin.

Once the spores of the Shaggy Ink Cap have been released, the gill tissue undergoes auto-digestion to form a black liquid that you can actually write with.

Right: Giant Puffballs can grow to be 70cm (27.6in) in diameter, this one shows damage caused by feeding slugs.

~ DRY-PRESERVING YOUR FUNGI ~

Life requires water, and the tiniest amount of water can create life in places where you might not want it, such as your specimen collection. Even in a mildly humid atmosphere a spore can germinate and turn meadowsweet into a mouldering mess, or a bacterium can breed and reduce a boletus to a blob. So it's best to keep things dry.

Preserving fungi presents its own set of problems – many will dissolve almost as soon as you look at them, others are juicy and more solid, and yet others have such fantastically complex and detailed structures that pressing will simply end in a nasty mess. For such specimens – and this works well with certain plant structures too – the best bet is to use a sand bath. The finer the sand, the better.

You will need:

- an ovenproof metal baking dish
- a baking tray and enough fine sand to cover it generously
- a fungus

1 Turn your oven to its highest setting, fill the baking dish with sand and bake it until it is as hot as you can make it – about half an hour. Carefully remove the tray using oven gloves.

2 Pour a layer of hot sand on to a baking tray, place your fungus on top of it and then pour in the rest of the sand, carefully brushing it into all crevices until the fungus is immersed in very hot sand.

3 Leave for a day before exhuming. If the fungus is not totally dry, repeat until it is. It is then ready for storage or presenting. Most specimens will be fine as they are, but to be on the safe side, store them in an airtight container with packets of desiccants such as silica gel, which can be purchased from electrical and photographic shops.

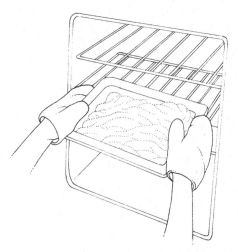

Step **1**

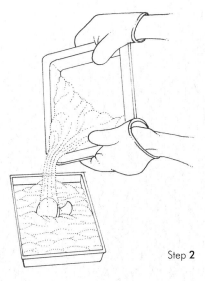

Step **2**

~ SPORE PRINTS ~

This is a rather fun way to see microscopic spores without a microscope and at the same time produce a beautiful representation of the elegant shapes of the gills and pores that release them from under the cap of a regular mushroom or toadstool.

You will need:

- a mushroom or toadstool
- a sharp knife
- a piece of black card
- a bowl large enough to fit over the top of the fungus

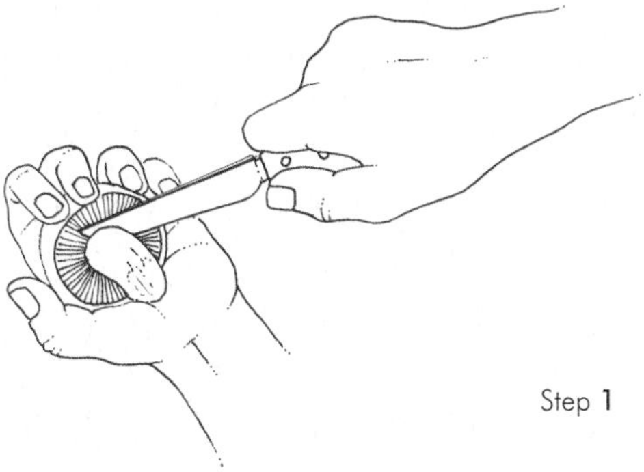

1 Cut the stem of the fungus off flush with the cap. Do this very gently and without touching the underside of the cap, as you might damage the fragile spore-releasing structures. Turn the cap the right way up and lay it, sporing surfaces down, on the card.

2 To stop air movements interfering with what you are trying to achieve, place a bowl over the mushroom – remember that spores are designed to be wafted by even the slightest breeze, and this will result in a less sharp impression.

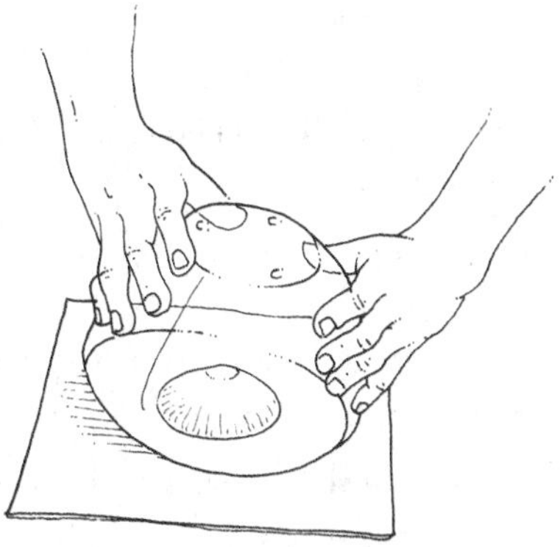

3 Leave for a day or so before carefully lifting first the bowl and then the fungus cap off the paper. You should be left with the pattern of the gills or pores, made by the millions of microscopic spores that have lain exactly where they have fallen. The pattern will soon be disrupted as the spores blow around, so if you want to preserve what you see, spray the print with fixative (available from art suppliers) or hairspray. This acts as a light glue, sticking everything together.

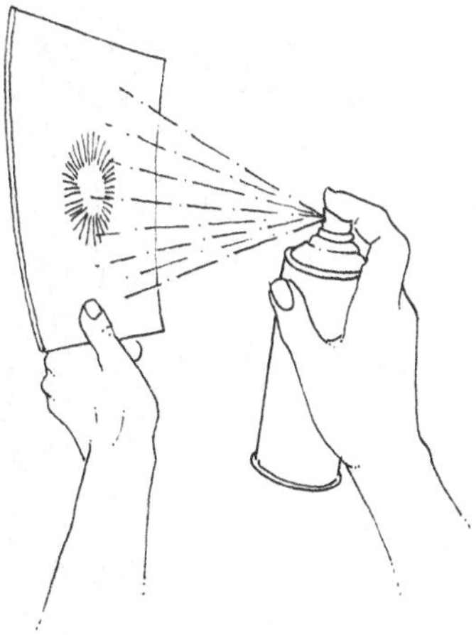

Step 1

Step 2

Step 3

Going further

I hope that this book will have sparked your interest and that you will want to go on to find out more information. One of the best ways of doing this is to contact some of the specialist societies. They usually have good websites with all sorts of links and information, and many produce leaflets and booklets, run conservation projects and will be able to tell you about events happening in your area. Joining a society can give you a chance to go on outings with experts in the area, and to receive useful information and newsletters. As well as the organisations listed here, remember there is probably a local natural history society or field club in your area.

Amphibian and Reptile Conservation Trust
655A Christchurch Road,
Boscombe,
Bournemouth,
Dorset,
BH1 4AP
Tel: +44 (0)1202 391319
Email: enquiries@arc-trust.org
www.arc-trust.org

The Amateur Entomologists Society and The Bug Club
PO Box 8774,
London,
SW7 5ZG
www.amentsoc.org

Botanical Society of the British Isles
Botany Department,
The Natural History Museum,
Cromwell Road,
London,
SW7 5BD
www.bsbi.org.uk

British Arachnological Society
www.britishspiders.org.uk

British Dragonfly Society
Conservation Officer,
c/o Natural England,
Parkside Court,
Hall Park Way,
Telford,
TF3 4LR
Tel: 0300 060 2338 (from inside UK)
www.british-dragonflies.org.uk

British Entomological and Natural History Society
The Secretary,
c/o The Pelham-Clinton Building,
Dinton Pastures Country Park,
Davis Street,
Hurst,
Reading,
Berkshire,
RG10 0TH
Email: enquiries@benhs.org.uk
www.benhs.org.uk

British Mycological Society
City View House,
5 Union Street,
Ardwick,
Manchester,
M12 4JD
Tel: +44 (0) 161 277 7638 / 7639
Email: admin@britmycolsoc.info
www.britmycolsoc.org.uk

British Naturalists' Association
General Secretary,
London,
WC1N 3XX
Tel: 0844 892 1817 (from inside UK)
www.bna-naturalists.org

British Trust for Ornithology
The Nunnery,
Thetford,
Norfolk,
IP24 2PU
Tel: +44 (0)1842 750050
Email: info@bto.org
www.bto.org

Buglife
Bug House,
Ham Lane,
Orton Waterville,
Peterborough,
PE2 5UU
Tel: +44 (0)1733 201 210
Email: info@buglife.org.uk
www.buglife.org.uk

Butterfly Conservation
Manor Yard,
East Lulworth,
Wareham,
Dorset,
BH20 5QP
Tel: +44 (0)1929 400209
Email: info@butterfly-conservation.org
www.butterfly-conservation.org

The Bat Conservation Trust
5th floor,
Quadrant House,
250 Kennington Lane,
London,
SE11 5RD
Tel: 0845 1300 228 (from inside UK)
Email: enquiries@bats.org.uk
www.bats.org.uk

The Bees, Wasps and Ants Recording Society
www.bwars.com

The Durrell Trust
Durrell Wildlife Conservation Trust,
Les Augrès Manor,
La Profonde Rue,
Trinity,
Jersey,
Channel Islands,
JE3 5BP
www.durrell.org

Field Studies Council
Head Office,
Preston Montford,
Montford Bridge,
Shrewsbury,
Shropshire,
SY4 1HW

Tel: 0845 3454071 (from inside UK)
Email: enquiries@field-studies-council.org
www.field-studies-council.org

Froglife
Head Office
1 Loxley,
Werrington,
Peterborough,
PE4 5BW
Tel: +44 (0)1733 602102
Email: info@froglife.org
www.froglife.org

Marine Conservation Society
Tel: +44 (0)1989 566017
www.mcsuk.org

The Mammal Society
3 The Carronades,
New Road,
Southampton,
SO14 0AA
Tel: +44 (0)2380 010981
E-mail: info@themammalsociety.org
www.mammal.org.uk

Plant Life
14 Rollestone Street
Salisbury
Wiltshire
SP1 1DX
Tel: +44 (0)1722 342730
www.plantlife.org.uk

Royal Society for the Protection Of Birds
The RSPB,
The Lodge,
Sandy,
Bedfordshire,
SG19 2DL
Tel: +44 (0)1767 693690
www.rspb.org.uk

Whale and Dolphin Conservation Society
Brookfield House,
38 St Paul Street,
Chippenham,
Wiltshire,
SN15 1LJ

Tel: +44 (0)1249 449500
Email: info@whales.org
http://uk.whales.org/

Wildfowl and Wetlands Trust
Tel: +44 (0)1453 891900
Email: enquiries@wwt.org.uk
www.wwt.org.uk

The Wildlife Trusts
The Kiln,
Mather Road,
Newark
United Kingdom
NG24 1WT
Tel: +44 (0)1636 677711
Email: enquiry@wildlifetrusts.org
www.wildlifetrusts.org/

The Woodland Trust
Kempton Way,
Grantham,
Lincolnshire,
NG31 6LL
Tel: +44 (0)1476 581111 / 581135
Email: england@woodlandtrust.org.uk

www.woodlandtrust.org.uk

Publications

Atropos magazine
www.atropos.info

Bird Watching magazine
www.greatmagazines.co.uk/bird-watching-magazine

Birdwatch magazine
www.birdwatch.co.uk

Birding World magazine
www.birdingworld.co.uk

British Birds magazine
www.britishbirds.co.uk

British Wildlife magazine
www.britishwildlife.com

Information

Natural England
www.naturalengland.org.uk

Countryside Commission for Wales
www.ccw.gov.uk

Scottish Natural Heritage
www.snh.gov.uk

Rare Bird Alert
www.rarebirdalert.co.uk

Discussion forums

RSPB Community
www.rspb.org.uk/community/forums

Birdforum
www.birdforum.net

Wild About Britain
www.wildaboutbritain.co.uk/forums/index.php

Further reading

The following titles are just a small selection of mainly European natural history books.

Angel, H. & Wolseley, P., *The Family Water Naturalist*, Michael Joseph (1982)

Arnold, E. N., *A Field Guide to the Reptiles and Amphibians of Britain and Europe*, Collins (2002).

Aulagnier, S. et al., *Mammals of Europe, North Africa and the Middle East*, A&C Black (2009)

Baker, N., *Nick Baker's Bug Book,* Bloomsbury (2015)

Baker, N., *Nick Baker's British Wildlife*, Bloomsbury (2015)

Baker, N., *The Nature Tracker's Handbook*, Bloomsbury (2013)

Bang, P. & Dahlstrom, P., *Animal Tracks and Signs*, Oxford University Press (2001)

Bellman, H., *A Field Guide to the Grasshoppers and Crickets of Britain and Northern Europe*, Collins (1985)

Blamey, M., Fitter, R. & Fitter, A., *Wild Flowers of Britain and Ireland*, 2nd edition. Bloomsbury (2013)

Brock, P. D., *A Comprehensive Guide to the Insects of Britain and Ireland*, Pisces Publications (2014)

Brooks, S. & Cham, S., *Field Guide to the Dragonflies and Damselflies of Great Britain and Ireland*, British Wildlife Publishing (2014)

Buczacki, S., *Collins Fungi Guide*, Collins (2013)

Burton, R. & Holden, P., *RSPB Birdfeeder Guide*, Dorling Kindersley (2003)

Carter, D. & Hargreaves, B., *A Field Guide to the Caterpillars of Britain and Europe*, Collins (1994)

Chinery, M., *A Field Guide to the Insects of Britain and Northern Europe*, Collins (1973)

Chinery, M., *Family Naturalist*, New Orchard Editions (1985)

Chinery, M., *Complete British Insects*, Collins (2005)

Chinery, M., *Insects of Britain and Western Europe*, Bloomsbury (2012)

Dietz, C. et al., *Bats of Britain, Europe & Northwest Africa*, A&C Black (2009)

Dijkstra, K., B., & Lewington, R., *Field Guide to the Dragonflies and Damselflies of Britain and Europe*, British Wildlife Publishing (2006)

Falkus, H., *Nature Detective*, Penguin (1982)

Fitter, R. & Fitter, A., *Grasses, Sedges, Rushes and Ferns of Britain and Northern Europe*, Collins (1984)

Fitter, R. & Manuel, R., *Field Guide to the Freshwater Life of Britain and North-West Europe*, Collins (1986)

Greenhalgh, M. & Ovenden, D., *Freshwater Life of Britain and Northern Europe*, Collins (2007)

Haahtela, T. et al., *Butterflies of Britain and Europe: A Photographic Guide*, A&C Black (2011)

Hammond, N., *The Wildlife Trusts Handbook of Garden Wildlife*, Bloomsbury (2014)

Harding, P., *How to Identify Edible Mushrooms,* Collins (1996)

Harrap, S., *Harrap's Wild Flowers*, Bloomsbury (2012)

Holden, P. & Abbott, G., *RSPB Handbook of Garden Wildlife*, Christopher Helm (2008)

Holden, P. & Cleeves, T., *RSPB Handbook of British Birds*, 4th edition, Bloomsbury (2014)

Lewington, R., *Guide to Garden Wildlife*, British Wildlife Publishing (2008)

Manley, C., *British Moths: Second Edition*, Bloomsbury (2015)

Roberts, M. J., *Spiders of Britain and Northern Europe*, Collins (1996)

Sterling, P. & Parsons, M., *Field Guide to the Micro-moths of Great Britain and Ireland*, British Wildlife Publishing (2012)

Thomas, A., *RSPB Gardening for Wildlife*, Bloomsbury (2010)

Tolman, T. & Lewington, R., *Collins Butterfly Guide*, Collins (2008)

Townsend, M. & Waring, P., *Field Guide to the Moths of Britain and Ireland,* British Wildlife Publishing (2009)

Bloomsbury and Collins produce an excellent range of field guides for most disciplines. There are numerous field guides available as phone Apps – some of the best are produced by Birdguides www.birdguides.com.

Image credits

Bloomsbury Publishing would like to thank the following for providing photographs and for permission to produce copyright material. While every effort has been made to trace and acknowledge all copyright holders, we would like to apologise for any errors or omissions and invite readers to inform us so that corrections can be made in any future editions of the book. As stated on the copyright page, all photos Nick Baker unless stated here below.

Photographs

Key t=top; l=left; r=right; tl=top left; tcl=top centre left; tc=top centre; tcr=top centre right; tr=top right; cl=centre left; c=centre; cr=centre right; b=bottom; bl=bottom left; bcl=bottom centre left; bc=bottom centre; bcr=bottom centre right; br=bottom right

FLPA Frank Lane Photography Agency; NPL: Nature Picture Library; G: Getty; SH: Shutterstock; Tony Cobley: Tony Cobley Photography, www.tonycobley.com

All photos by Nick Baker unless specified as below.

Front cover: Tony Cobley, title lettering/spine SH; **back cover:** Tony Cobley, top row SH; **half-title:** Tony Cobley; **6** Jan Vermeer/Minden Pictures/FLPA; **8** Tony Cobley; **12** Tony Cobley; **13** Tony Cobley; **15** Tony Cobley; **p21** Terry Whittaker/2020VISION/NPL; **22** SH; **25** Bloomsbury; **28** t Tony Cobley; **30** Terry Whittaker / 2020VISION; **31** Ross Hoddinott / 2020VISION/NPL; **32** tl SH, tr SH, c Albert Lleal/Minden Pictures/FLPA, b SH; **33** tr Tony Cobley; **34** Tony Cobley; **36** b Cynthia Monaghan / Contributor/G; **37** tr Tony Cobley; **38** Tony Cobley; **39** SH; **40** Robin Chittenden/FLPA; **41** SH; **42** SH; **p44** t Chris Shields (WAC)/NPL, b Paul Sawer/FLPA; **45** t , bl , br SH; **47** t Jean-Lou Zimmerman/Biosphoto/FLPA, b Paul Hobson/FLPA; **48** t SH, b Andy Rouse/NPL; **50** SH; **51** t Do Van Dijck/Minden Pictures/FLPA, tc Duncan Usher/Minden Pictures/FLPA, br Duncan Usher/Minden Pictures/FLPA; **52** t, b SH; **53** Paul Sawer/FLPA; **54** Hugh Pearson; **55** Tony Cobley; **56** SH; **57** David Tipling/FLPA; **58** c, b SH; **60** t David Tipling/FLPA, b SH; **61** t, b SH; **62** t SH, b Gianpiero Ferrari/FLPA; **63** c SH; **64** t SH, b Erica Olsen/FLPA; **67** t Roger Wilmshurst/FLPA, b SH; **68** t, b SH; **69** b SH; **70** t Ingo Arndt/Minden Pictures/FLPA, b David Hosking/FLPA; **71** t Richard Du Toit/Minden Pictures/FLPA, b SH; **72** SH; **73** t, b SH; **74** SH; **75** t Duncan Usher/Minden Pictures/FLPA, b Andrew Parkinson/NPL; **76** SH; **77** Robert Canis/FLPA; **78** t SH; **79** t Michael Clark/FLPA, b SH; **81** t SH, b Niall Benvie; **82** Derek Middleton/FLPA; **83** t, b SH; **84** SH; **86** Pete Oxford/Minden Pictures/FLPA; **88** Tony Heald/NPL; **91** Andrew Mason/FLPA; **92** Martin B Withers/FLPA; **96** t David Hosking/FLPA, b Robert Canis/FLPA; **98** Martin Camm (WAC)/NPL; **103** t David Hosking/FLPA, b Nick Upton / 2020VISION; **106** SH; **107** t, b SH, **108** Terry Whittaker / 2020VISION; **109** SH; **110** SH; **111** SH; **114** all SH, br Photo Researchers/FLPA; **116** tr, br SH; **117** A© Biosphoto, David Massemin /FLPA; **119** tr SH, br Andrew Murray; **121** SH; **122** t Matt Cole/FLPA, br ImageBroker/Imagebroker/FLPA; **123** t Fabio Liverani, br Emanuele Biggi/FLPA; **124** SH; **125** tr Robert Henno/Biosphoto/FLPA, br Jack Perks/FLPA; **127** SH, Loic Poidevin/NPL; **128** Photo Researchers/FLPA; **130** t Imagebroker, Christian Hatter/Imagebroker/FLPA; **131** Paul Hobson/FLPA; **133** Guy Edwardes/NPL; **134** SH; **136** Michel Gunther/Biosphoto/FLPA; **137** Sandesh Kadur/NPL; **138** Alex Hyde/NPL; **139** SH; **140** MYN/Carsten Krieger/NPL; **144** Chris Brignell/FLPA; **147** br Ingo Arndt/Minden Pictures/FLPA; **148** OceanPhoto/FLPA; **149** D P Wilson/FLPA; **150** t Linda Pitkin/2020VISION/NPL, b SH; **151** Kim Taylor/NPL; **152** Alan James/NPL; **153** t Dan Burton/NPL, b David Tipling/NPL; **154** Graham Eaton/NPL; **155** tr Juan-Carlos Munoz/Biosphoto/FLPA, b Linda Pitkin/2020VISION/NPL; **157** Oriol Alamany/NPL; **158** SH; **160** Tim Laman/NPL; **161** tr Remi Masson/Biosphoto/FLPA, br SH; **162** SH; **163** SH; **165** tl, tr SH, bl Otto Plantema/Minden Pictures/

FLPA, br Piotr Naskrecki/Minden Pictures/FLPA; **166** tl Steve Trewhella/FLPA, bl Piotr Naskrecki/Minden Pictures/ FLPA; **167** SH; **168** b Nigel Cattlin/FLPA; **172** t Tony Cobley, b Heidi & Hans-Juergen Koch/Minden Pictures/FLPA; **173** SH; **174** Tony Cobley; **175** SH; **176** Albert Lleal/Minden Pictures/FLPA; **177** tr Lars Soerink/Minden Pictures/ FLPA; **178** Chien Lee/Minden Pictures/FLPA; **179** SH; **181** SH; **182** tr SH, b Thomas Marent/Minden Pictures/FLPA; **183** t Kim Taylor/NPL, br Gary K Smith/FLPA; **184** tr Kim Taylor/NPL, c Hans Christoph Kappel/NPL, br SH; **185** tr SH, br Simon Colmer/NPL; **186** tl Wild Wonders of Europe/Zupanc/NPL, br John van den Heuvel/Minden Pictures/ FLPA; **187** tr SH, bl Ingo Arndt/NPL; **188** tr SH, bl Malcolm Schuyl/FLPA; **189** tr Robert Thompson/NPL, c, br David M. Cotteridge; **190** t David M. Cotteridge; **191** David T. Grewcock/FLPA; **192** Stephen Dalton/NPL; **195** John Walters; **196** Peter Entwistle/FLPA; **197** cl, br SH; **198** tl Thomas Marent/Minden Pictures/FLPA, c SH; **199** tr Rene Krekels, NiS/Minden Pictures/FLPA, br Jan Hamrsky/NPL; **201** tr Derek Middleton/FLPA, br SH; **203** SH; **204** Nick Upton/NPL; **205** tl Cisca Castelijns, NiS/Minden Pictures/FLPA, br Premaphotos; **206** SH; **207** t SH; **208** SH; **209** t, b SH; **210** tl, tr, cl, cr, br SH, bl ©Biosphoto, Robert Henno/Biosphoto/FLPA; **211** tr, bl SH; **213** t Paul Hobson/NPL, br Michael Durham/NPL; **214** Stephen Dalton/NPL; **215** tr, bl SH; **216** SH; **217** tl Visuals Unlimited/NPL, c John B Free/NPL, br SH; **218** tr Adrian Davies/NPL br Imagebroker, Franz Christoph Robi/Imagebroker /FLPA; **219** tl John B Free/NPL, c Mark Moffett/Minden Pictures/FLPA, br Bernard Castelein/NPL; **220** Richard Becker/FLPA; **226** t SH; **227** tr Andy Sands/NPL, br SH; **228** tr Visuals Unlimited/NPL, b Ingo Arndt/Minden Pictures/FLPA; **230** bl Richard Becker/FLPA; **232** tr, b, SH; **233** t, b SH; **235** Premaphotos/NPL; **236** SH; **237** t Matt Cole/FLPA, b ; **238** Jurgen Freund/NPL; **239** t SH, b Imagebroker, J.Walker/Imagebroker/FLPA; **240** Nigel Cattlin/FLPA; **241** t SH; **244** t Olivier Digoit/Imagebroker/FLPA, b Emanuele Biggi/FLPA; **245** Michael Durham/Minden Pictures/FLPA; **247** ImageBroker/ Imagebroker/FLPA; **248** Steve Trewhella/FLPA; **249** t, br SH; **250** Jan Hamrsky/NPL; **252** Nicholas and Sherry Lu Aldridge/FLPA; **253** t, c SH; **254** b Wil Meinderts/Minden Pictures/FLPA; **255** t Steve Trewhella/FLPA, b Kim Taylor; **256** t Wikimedia/ Hans Hillewaert, b Steve Trewhella/FLPA; **257** t Steve Trewhella/FLPA, b SH; **258** Jack Perks/FLPA; **259** t, bl, br SH; **260** t, b SH; **261** SH; **262** Tony Hamblin/FLPA; **263** tr SH, br Kim Taylor/NPL; **264** Jane Burton/ Minden Pictures/FLPA; **265** tr SH, br Nick Upton/NPL; **266** tl Christoph Becker/NPL, c John Cancalosi/NPL; **267** tl Ingo Arndt/Minden Pictures/FLPA, c Simon Colmer/NPL, bl SH; **268** t Nick Upton/NPL, br Tony Hamblin/FLPA; **269** tr Alex Mustard/NPL, bl Colin Marshall/FLPA; **270** t Steve Trewhella/FLPA, c SH, br Bill Coster/FLPA; **272** tr Steve Trewhella/FLPA, br Willem Kolvoort/NPL; **273** tr Nick Upton/NPL, br Malcolm Schuyl/FLPA; **274** tr Steve Trewhella/ FLPA, br Steve Trewhella/FLPA; **275** tl D P Wilson/FLPA, br Christophe Courteau/NPL; **276** tr Visuals Unlimited/NPL, bl SH; **277** tr Philippe Clement/NPL, br SH; **278** tl ImageBroker/Imagebroker/FLPA, br Steve Trewhella/FLPA; **279** br Chris Mattison/FLPA; **280** Nigel Cattlin/FLPA, Nick Upton/NPL; **281** tr, bl SH; **282** br SH; **283** Derek Middleton/ FLPA, **284** SH; **286** ©Biosphoto, Sylvain Cordier/Biosphoto/FLPA; **287** tr D P Wilson/FLPA, br Pete Oxford/Minden Pictures/FLPA; **288** tl, bl Steve Trewhella/FLPA; **289** Jabruson/NPL; **290** SH; **291** SH; **293** tr Nigel Cattlin/FLPA, br Jurgen Freund/NPL; **294** bl, br SH; **295** Nigel Cattlin/FLPA; **296** cr, lr SH; **298** cr Visuals Unlimited/NPL, bl Bob Gibbons/FLPA; **301** cr, bl SH, **302** tr, br SH; **303** tr SH, br Chien Lee/Minden Pictures/FLPA; **306** Kurt Möbus/ Imagebroker/FLPA; **308** SH; **309** t Duncan McEwan/NPL; **310** Kim Taylor/NPL; **312** SH; **313** tr, bl SH; **314** Erica Olsen/FLPA; **315** cr, br SH; **316** Stephen Dalton/NPL; **320** tr Jabruson/NPL, br SH; **321** tr Jurgen & Christine Sohns/ FLPA, br SH; **325** Tony Cobley; **326** Premaphotos/NPL; **327** Visuals Unlimited/NPL; **328** SH; **329** tr SH, bl Jurgen Freund; **331** tr Nico van Kappel/Minden Pictures/FLPA, bl Nigel Bean/NPL; **332** tl Steve Gettle/Minden Pictures/ FLPA, cr SH; **333** Stephen Dalton/NPL

Illustrations by Lizzie Harper, except for p44 Dave Nurney, p95 and 98 Wildlife Art Company/NPL

Index